# Graphene and other Two-dimensional Materials in Nanoelectronics and Optoelectronics

# Graphene and other Two-dimensional Materials in Nanoelectronics and Optoelectronics

Special Issue Editor

**Jie Sun**

MDPI • Basel • Beijing • Wuhan • Barcelona • Belgrade • Manchester • Tokyo • Cluj • Tianjin

*Special Issue Editor*
Jie Sun
Fujian Science & Technology Innovation Laboratory for Optoelectronic Information of China,
and Fuzhou University
Sweden

*Editorial Office*
MDPI
St. Alban-Anlage 66
4052 Basel, Switzerland

This is a reprint of articles from the Special Issue published online in the open access journal *Materials* (ISSN 1996-1944) (available at: https://www.mdpi.com/journal/materials/special_issues/graphene_2d).

For citation purposes, cite each article independently as indicated on the article page online and as indicated below:

LastName, A.A.; LastName, B.B.; LastName, C.C. Article Title. *Journal Name* **Year**, *Article Number*, Page Range.

**ISBN 978-3-03936-204-2 (Hbk)**
**ISBN 978-3-03936-205-9 (PDF)**

# Contents

# About the Special Issue Editor

**Jie Sun** got his Bachelor's and Master's degrees in microelectronics and solid-state electronics from the Dalian University of Technology. He has a PhD in semiconductor materials from the Institute of Semiconductors, Chinese Academy of Sciences, and another PhD in semiconductor devices from the Solid State Physics Division, Lund University. He is now based at Fujian Science & Technology Innovation Laboratory for Optoelectronic Information of China, and Fuzhou University, conducting research on GaN-based microLEDs as well as 2D materials and devices. He is a senior member of the IEEE. He has published 111 papers (Web of Science) and has an H-index of 24. He has been granted seven patents.

# Preface to "Graphene and other Two-dimensional Materials in Nanoelectronics and Optoelectronics"

I started out as a Master's student working on LPE of GaAs for LEDs and lasers in 2000. It was already an old-fashioned material technology back then, and it was not easy to obtain good devices based on those GaAs. However, I did learn a truth: no good materials, no good devices. Soon after this, I had the opportunity to access more advanced epitaxial techniques (e.g., MBE and CVD) which fortified my belief that novel and decent materials are the ultimate limiting factor on the advancement of solid state electronics.

Graphene has been known to mankind for a long time, but it was not until 2004, when it was first isolated on $SiO_2$ and its field effect measured, that people understood its revolutionary role in nanoelectronics. Indeed, there had never been any other material simultaneously possessing such outstanding properties, including high mobility, high transparency, high thermal conductivity, high mechanical strength and flexibility. Thus, graphene was given the name "wonder material". Today, there are several ways to produce it, such as micromechanical exfoliation, chemical vapor deposition, etc. In the first article of this book, the authors introduce a new method—glow discharge—to synthesize graphene. Although glow discharge is an old concept, it is "new" here, in the sense that it is the first time it has been used to grow thin film graphene as opposed to flake graphene, which is suitable for device applications.

When I pursued my PhD, I switched from being a material scientist to being a device physicist and engineer, and I spent four years in front of an electron beam lithography machine fabricating InP-based nanoelectronic devices. There I understood another truth: any good material should be compatible with processing devices, otherwise it will be literally useless in applications. Regarding graphene, one of its most obvious advantages over other nanomaterials is that it is two-dimensional, which is compatible with existing semiconductor fabrication. Indeed, in the second paper of this book, the authors show that graphene can be integrated onto GaN wafers as the transparent electrode of microLEDs and the transistor channels of their drivers. This application has a great deal of potential. Graphene is transparent and flexible, which is suitable for transparent and flexible microLED-based displays (with sapphire substrate removed). Nevertheless, graphene processing today also has a bottleneck: the difficulty of transfer. Graphene is typically grown by CVD on a foreign substrate and needs to be mechanically transferred. Holes, wrinkles, and contaminations are inevitable. The third paper offers a solution: to grow graphene directly on GaN and use it in situ as the transparent electrodes for LEDs. It is challenging to obtain high crystalline quality for this direct growth method, but it seems we have to live with that—unless we can accept graphene transfer, which we cannot for real-world applications.

Papers four and five in this book show it is quite possible to go in the opposite direction. That is, to prepare nitride semiconductor materials/devices on graphene and other 2D materials. This is a promising new direction, where 2D materials could potentially serve as cheap/flexible substrates for nitrides. It is worth mentioning that $WS_2$ has a bandgap, meaning it is a real semiconductor. Of course, it has an advantage, rendering it superior to graphene for fabrication of many electronic devices. If Si, GaAs, and GaN represent first-, second-, and third- generation semiconductors, respectively, I would argue that 2D semiconductors may constitute the fourth generation.

Finally, the thermal aspects of 2D materials are relatively overlooked; namely, they seem to have received less attention than the optical/electrical properties. However, thermal management is vital in today's electronics, as integrated electron devices are increasingly densely packed. Indeed, graphene has been suggested as a heat spreader in electronic chips. Paper six, however, demonstrates the opposite application—graphene can be used as a transparent heater, where a convective heat-transfer coefficient of 60 W/(°Cm$^2$) is achieved, better than many other heaters. Of course, this is based on the unique thermal properties of graphene and other 2D materials. The seventh paper reviews a resistance thermometer for evaluating the thermal properties of low-dimensional materials, which might be useful for cooling electronic chips.

The authors of these articles are students and professors working actively on semiconductors and 2D materials and devices. Some of these are experts I have already known for two decades. I am impressed by their work and scientific attitude, and I am grateful to them for making the publication of this book possible. An old saying goes: "cast a brick to attract jade". Likewise, I hope this book will inspire new and innovative ideas in the fascinating field of 2D-materials-based nanoelectronics and optoelectronics.

**Jie Sun**
*Special Issue Editor*

Article

# High Quality Graphene Thin Films Synthesized by Glow Discharge Method in A Chemical Vapor Deposition System Using Solid Carbon Source

**Le Wang [1], Jie Sun [2,3,*], Weiling Guo [1,†], Yibo Dong [1], Yiyang Xie [1], Fangzhu Xiong [1], Zaifa Du [1], Longfei Li [1], Jun Deng [1] and Chen Xu [1]**

1 Key Laboratory of Optoelectronics Technology, Beijing University of Technology, Beijing 100124, China; wangle316@emails.bjut.edu.cn (L.W.); guoweiling@bjut.edu.cn (W.G.); donyibo@emails.bjut.edu.cn (Y.D.); xieyiyang@bjut.edu.cn (Y.X.); fangzhuxiong@emails.bjut.edu.cn (F.X.); 17801011216@163.com (Z.D.); 18800153166@163.com (L.L.); dengsu@bjut.edu.cn (J.D.); xuchen58@bjut.edu.cn (C.X.)

2 National and Local United Engineering Laboratory of Flat Panel Display Technology, College of Physics and Information Engineering, Fuzhou University, Fuzhou 350100, China

3 Fujian Science & Technology Innovation Laboratory for Optoelectronic Information of China, Fuzhou 350100, China

* Correspondence: jie.sun@fzu.edu.cn

† Contributed equally to this work.

Received: 19 February 2020; Accepted: 8 April 2020; Published: 26 April 2020

**Abstract:** Arc discharge is traditionally used to synthesize randomly arranged graphene flakes. In this paper, we substantially modify it into a glow discharge method so that the discharge current is much more reduced. The $H_2$ and/or Ar plasma etching of the graphitic electrode (used to ignite the plasma) is hence much gentler, rendering it possible to grow graphene in thin film format. During the growth at a few mbar, there is no external carbon gas precursor introduced. The carbon atoms and/or carbon containing particles as a result of the plasma etching are emitted in the chamber, some of which undergo gas phase scattering and deposit onto the metallic catalyst substrates (Cu-Ni alloy thin films or Cu foils) as graphene sheets. It is found that high quality monolayer graphene can be synthesized on Cu foil at 900 °C. On Cu-Ni, under the same growth condition, somewhat more bilayer regions are observed. It is observed that the material quality is almost indifferent to the gas ratios, which makes the optimization of the deposition process relatively easy. Detailed study on the deposition procedure and the material characterization have been carried out. This work reveals the possibility of producing thin film graphene by a gas discharge based process, not only from fundamental point of view, but it also provides an alternative technique other than standard chemical vapor deposition to synthesize graphene that is compatible with the semiconductor planar process. As the process uses solid graphite as a source material that is rich in the crust, it is a facile and relatively cheap method to obtain high quality graphene thin films in this respect.

**Keywords:** graphene; glow discharge; graphite; chemical vapor deposition; metal catalyst; solid carbon source; plasma

---

## 1. Introduction

Since its appearance, graphene has been applied to solar cells, sensors, composite materials, photocatalysis, and other fields due to its outstanding characteristics such as high carrier mobility, high mechanical strength, high specific surface area, high transmittance, and high thermal conductivity [1]. The graphene synthesis methods include mechanical exfoliation, chemical exfoliation, epitaxial graphene on SiC [2], molecular beam epitaxy (MBE) [3], chemical vapor deposition (CVD) [4], and arc discharge (AD) [5–13], etc. Among these technologies, mechanical and chemical exfoliation of graphite can only produce irregular shaped flakes or powders; epitaxial graphene on SiC is hardly transferable, very expensive, and the graphene film size is limited by the available SiC substrate; MBE graphene is not mature and the material quality is modest; only CVD technique can produce scalable graphene thin films that are compatible with standard semiconductor planar process. Regarding AD, because of the low controllability, it has not received widespread attention in graphene synthesis thus far. Traditionally, the AD method is used to synthesize fullerenes, single or multi-wall carbon nanotubes, randomly arranged graphene flakes, carbon nanoparticle-based light-emitting devices, etc. [5–13]. When making graphene by AD, the graphite electrode is usually used as the carbon source. Typically, the electrode can be completely consumed in just ten minutes or so due to the high electrical current and the intense etching reaction, and the graphene grown is multi-layered (about 2 to 10 layers) and very defective [5–13]. A variant version of the AD method is the synthesis of graphene by the so-called hydrogen AD exfoliation of graphite, which is often with the involvement of graphene oxides [13]. Nevertheless, it is also not very controllable and, most importantly, it is not compatible with today's semiconductor processing and cannot produce graphene in the format of thin films.

In this paper, we have substantially modified the traditional AD method, based on which we introduce a new graphene synthesis technique that is called the glow discharge (GD) deposition method. The equipment used in this work is actually a standard PECVD (plasma enhanced CVD) system that was not originally intended for the GD use. In the chamber, which is filled with $H_2$ and/or Ar gas, we ignited the plasma, and it slightly etches the graphite electrode that is used to start the plasma. Because of this, we were able to grow graphene on the surface of metal catalysts even without using any carbon precursor gas. The possible deposition mechanism is explained by the plasma-based physical and/or chemical etching of the graphitic electrode, followed by gas phase scattering of the produced carbon atoms and carbon based particles, which are in part transported towards the catalytic surfaces of the metallic substrates situated on the heater. After the catalytic graphitization on the metal surface, the prepared graphene/metal is unloaded, and the graphene can be transferred to insulators using a wet etching-based technique [14,15]. The parameters of the GD method are systematically optimized, and the materials are characterized in detail. Compared with MBE, partly due to the existence of catalyst, the GD offers a better material quality, and the growth time is also much shorter. Even including the pumping and heating/cooling procedure, it only takes 20 min for one run. Compared to standard CVD, the GD offers an alternative method that negates the need for work on the precursor gas composition, the decomposition rate, and the large growth parameter space. Furthermore, the growth temperature of GD is somewhat lower. Compared to AD, the GD here is a much gentler process. Most importantly, the graphene prepared by GD is a continuous thin film with high quality, which can be transferred to insulators and used to make electronic devices. Our work explores the potential of the traditionally overlooked AD method in synthesizing graphene thin films that are compatible with semiconductor planar processing. Therefore, it is of value to scientists and engineers who work with the synthesis of graphene and its electronic device applications.

## 2. Experimental Procedures and Methods

In order to verify and benchmark the quality of the graphene obtained by the new GD graphene production mode proposed in this paper, we first grow standard graphene thin films by CVD using two kinds of conventional catalytic metals (Cu-Ni alloy film and copper foil) [16,17] as control samples. The CVD process is described in our earlier publication [14,15]. Those two types of metals are also the substrates that are used to grow graphene by our GD method, and their preparation methods are as follows.

(1) Cu-Ni alloy thin films. After the Si wafer is cleaned by standard procedure, 300 nm $SiO_2$ is grown by inductively coupled plasma chemical vapor deposition (ICP-CVD). This is because, later, the Cu-Ni will be removed and the grown graphene will "land" on the $SiO_2$, and 300 nm silicon dioxide is known to facilitate the observation of graphene under an optical microscope after growth, due to the

optical interference effect [18]. Magnetron sputtering is used to sputter Cu-Ni (2:1) alloys of different thicknesses on the $SiO_2$/Si wafers to study the effect of catalyst thickness on the quality of graphene.

(2) Copper foils. High purity polycrystalline copper foils are purchased commercially.

The growth system used in this work is a graphene PECVD of the model Black Magic, produced by Aixtron Nanoinstruments Ltd. The outwall of the growth chamber is made of stainless steel with a quartz shield as the inwall. As shown in Figure 1, gases come from the top via a quartz showerhead, which has small holes to distribute the gases uniformly to the samples below. The arrows indicate the direction of the gas flow. The samples are placed on a graphitic heater supported by two vertical metal rods which are also working as the heater electrodes. Alternating current (AC) current is sent through the two electrodes to heat the heater by Joule heating. On the heater, there is a ceramic clamp which can hold a maximum 2 inch sample. A third electrode, which we call the plasma electrode, is right below the heater (with ~4 cm distance). An AC or direct current (DC) voltage of several hundred volts can be applied to the plasma electrode, with respect to the potential of the heater in order to ignite the plasma (In this paper we have used the AC configuration).

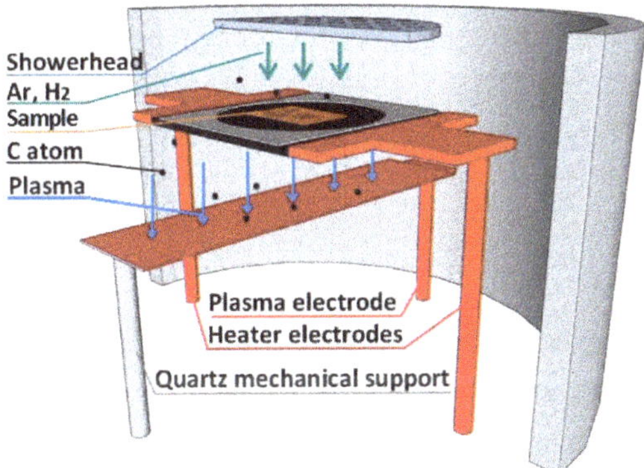


**Figure 1.** Schematic illustration of the growth chamber of the plasma enhanced chemical vapor deposition (PECVD) equipment which is used for the glow discharge (GD) synthesis of graphene in this work.

We place the as-prepared Cu-Ni alloy thin film or copper foil samples on the graphite plate heater in the equipment (see Figure 1) for the GD growth of graphene. The copper foil sample, however, should be first put in a quartz bowl, because if the foil is in direct contact with the heater, the electrical current leaked into the copper from the graphite is hard to control, which will add extra heating and may melt the copper. After evacuating to a base pressure of $2 \times 10^{-3}$ kPa, we introduce Ar and/or $H_2$ gases (for example, with a ratio of 5:1) into the chamber, and the pressure reaches $5 \times 10^{-1}$ kPa. The procedure is repeated three times and then the gas input is turned off. This is because we have found that without a flowing gas, the plasma power during the initiation is more stable than is the case with a gas flow. The heating rate is 5 °C/s, at which we increase the heater temperature to the desired growth temperature (typically 600–1000 °C). The samples are held at this temperature for ten minutes for annealing to increase the crystallinity of the metals [19], which is especially useful for the Cu-Ni alloy. Then, the plasma is ignited. The power intensity is 40 W, and the AC frequency is 20 kHz. After 5 min of growth, the plasma is turned off, and the temperature is cooled down to 500 °C at a rate of 5 °C/s. Afterwards, the Ar gas is introduced to the chamber to accelerate the temperature dropping until it is cooled to room temperature, and the samples are unloaded.

After growth, the copper-nickel alloy and the copper foil samples are all spin-coated (4000 rpm for 30 s) with a layer of polymethyl methacrylate (PMMA) with about 70 nm thickness. Then, it is heated on a 150 °C hot plate for 10 min. A metal etching solution ($CuSO_4$:HCl:$H_2O$ = 5 g:25 mL:50 mL) is prepared. The copper-nickel sample is placed at the bottom of the beaker, whereas the copper foil sample is floating on the surface of the solution. The copper foil grown graphene is transferred to another 300 nm $SiO_2$/Si substrate by the standard wet transfer process [14,15]. The graphene grown on the Cu-Ni alloys, however, is transferred to its own $SiO_2$/Si substrate through the etching mechanism that is shown in Figure 2. The chemical solution penetrates the PMMA and etches the metal beneath. Because of the buffering of the PMMA layer, the reaction will become very gentle, which reduces the possibility of graphene being damaged by the etching process [17]. After etching, we put the PMMA/graphene/$SiO_2$/Si complex in deionized water for 30 min to clean up the residual chemicals. In order to improve the adhesion between the graphene and the substrate, it is baked at 150 °C for 15 min on a hot plate. Then, it is placed in acetone for 1 h to remove the PMMA from graphene. Finally, the sample is placed in a ventilated place for 10 min to dry the acetone. During the experiment, nevertheless, after the metal etching the van der Waals force between the $SiO_2$/Si substrate and the PMMA/graphene is relatively weak, and the rinsing process in deionized water might cause the PMMA/graphene to float on the surface of solution. Applying a little PMMA to the edges of the sample helps hold the film, and can increase the success rate of the graphene transfer.

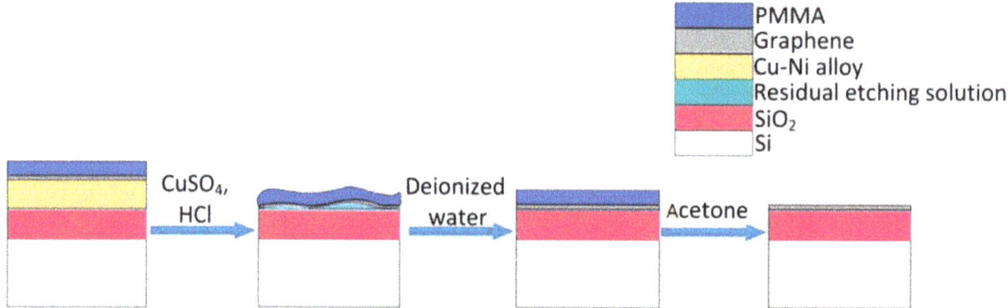


**Figure 2.** Process of "transferring" the graphene grown by Cu-Ni alloy onto its own $SiO_2$/Si substrate.

After the sample preparation, we use Raman spectroscopy with the excitation wavelength at 532 nm to characterize the quality of the graphene thin films, as well as to determine the number of graphene layers grown with different metal catalysts. Raman mapping is also performed. A scanning electron microscope (SEM, MERLIN Compact, Zeiss, Oberkochen, Germany) is used to examine the morphology of the metals after the annealing and after the graphene deposition.

## 3. Results and Discussion

The schematic illustration of the graphene deposition procedure is shown in Figure 3a. The production mechanism of graphene involves two steps: (1) the creation of carbon atoms or carbon containing particles; (2) the graphitization of these particles. Regarding the graphitization on the substrate, it is the same for both the GD growth method and the normal graphene CVD. However, in a regular CVD, the formation of carbon atoms or carbon containing particles is via the decomposition of hydrocarbons, whereas in GD it is through the gentle plasma etching of the electrode. The possible carbon particle formation mechanism of our GD method is explained as follows. GD and AD are two typical discharge procedures of gaseous species. Compared to the traditional AD method, the process in our GD is much gentler. Usually, GD occurs at a higher and more stable voltage, but the current is smaller (in mA). AD happens at lower voltage (typically 1~10 V), but the current is in ampere range (typically 10~100 A). It's often much brighter and hotter compared to GD. In our machine, the applied plasma voltage is in the order of several 100 volts, the current is 0.1–0.5 A, and the pressure is a few mbar. These are typical conditions for GD but not for AD [20]. Under our condition, therefore, the phenomenon should be defined as GD instead of AD, which is a lot gentler. The plasma etches the graphite, but does not exfoliate it like in the AD. The emitted carbon atoms and carbon containing particles are graphitized into textured thin films on the metal surfaces. That eliminates the possibility to produce random graphene flakes as the AD method does. We have not found any reports about graphene thin films produced by AD. Therefore, the graphene produced by AD cannot be applied in electronic chips. It is mainly for other applications such as composite materials. In our GD, on the other hand, as shown later, no matter what Ar to $H_2$ gas ratios are used, we can always obtain thin film graphene conformally coating on the metal catalysts, provided that other conditions are optimized (e.g., temperature, pressure, plasma power). Since there is no carbon containing gas introduced in the machine, the only explanation is that the graphene grows from the etched graphite. Note that we have found the graphene thin films do not form when no plasma is present. Thus, we can conclusively exclude the possibility that the graphene's carbon source comes from unintentionally introduced carbon species in the chamber such as carbon contamination, oil vapor from the pump, etc. In our experiment, the etching mechanisms of $H_2$ and Ar plasmas are different. The former ($H_2$ plasma) is mainly a chemical process, where the etching of graphite is achieved through hydrogenation (forming C-H bonds) and the subsequent releasing of hydrocarbons such as $CH_4$ [21]. Some authors, however, suggest a slightly different mechanism, where the hydrocarbon formation is a result of first ionic bombardment and subsequent chemical reaction [22]. The latter (Ar plasma) is basically a physical process, where the etching is simply an ionic bombardment and creates carbon atoms [23,24]. So, how can those carbon atoms or carbon-based particles that resulted from the etched graphite turn into graphene thin films? When the plasma electrode is etched, the produced particles are transported onto the catalytic metal substrates via gas phase scattering. The mean free path $\lambda$ of the particles can be approximately estimated by the ideal gas formula $\lambda = \frac{1}{\sqrt{2}\pi d^2 n}$, where $d$ is the effective diameter of the particle, and $n$ is the number of particles per unit volume. Using that equation, the mean free path of the carbon-based particles is in the order of a few millimeters, which is much shorter than the distance between the plasma electrode and the samples on the heater (~4 cm). Therefore, the particles undergo many scatterings in the chamber and some particles will have chances to be directed toward the samples, as depicted in Figure 3a. In fact, since the graphite heater is also immersed in the plasma, it can be slightly etched, and supply some of the carbon particles as well. Because of the existence of the catalysts, graphitization happens on their flat surfaces and graphene thin films grow therein.

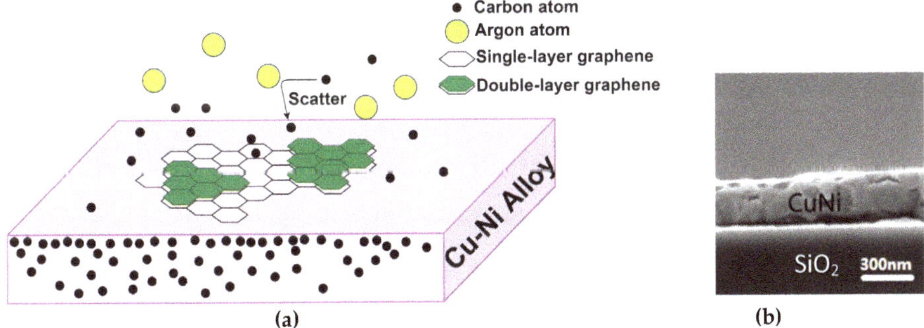


**Figure 3.** (**a**) Schematic diagram of the deposition of graphene by GD method on the Cu-Ni alloy. The carbon atoms resulted from the plasma etching of the graphite electrode are transported to the metal catalytic surface via scattering. (**b**) Scanning electron microscope (SEM) image (cross-sectional view) of the Cu-Ni alloy after annealing at 900 °C.

The introduction of the GD production mode in the PECVD chamber makes it feasible to grow thin film graphene under our conditions, which is not possible in the traditional AD method, which can only produce chaotic flakes. That is a new feature in graphene synthesis by gas phase discharge, both from fundamental and application points of view. Furthermore, in the GD method, the graphene is grown from cheap graphite and not an expensive carbon precursor gas, reducing the cost in this regard. Also, because there is no carbon precursor gas, it makes the process optimization much easier, without the need to monitor the precursor gas ratios. Only $H_2$ and/or Ar gases are required in the chamber. As can be seen later, the quality of graphene is insensitive to the $H_2$ to Ar ratio. According to our experiment, the material quality can be optimized by adjusting parameters such as the type of catalytic metals and the deposition temperature. Finally, after the GD growth, we have measured the 300 μm thick plasma electrode and could not detect any weight loss. Therefore, the process is indeed very gentle, and the consumption of the graphite material is tiny. In our practice, the same graphite electrode can be used for hundreds of runs.

In Figure 3b, the cross-sectional morphology of the Cu-Ni alloy after annealing can be seen. The Cu-Ni thin film thickness is 300 nm. It can be seen that the film is rather flat, and the copper and nickel have been uniformly alloyed after the annealing at 900 °C. The Raman results of the Cu-Ni alloy grown graphene samples can be seen in Figure 4a. The G band appears in each curve as a typical signature of a $sp^2$ hybridized graphitic carbon structure. As the temperature rises from 600 °C to 1000 °C, the characteristic 2D peaks of the graphene Raman spectroscopy gradually appear. Meanwhile, the D peak eventually decreases. These features indicate that the number of defects or disorders in the graphene decreases. The Raman characterization results of the samples at 800 °C–1000 °C are more or less similar to each other. The Raman results are also comparable with standard CVD graphene grown in the same type of machine (see our previous publications [14,15]), which confirms the crystalline quality of the as-grown graphene. The feasibility of using the GD method to grow reasonably high-quality graphene with catalyst at around 900 °C is thus experimentally proved.

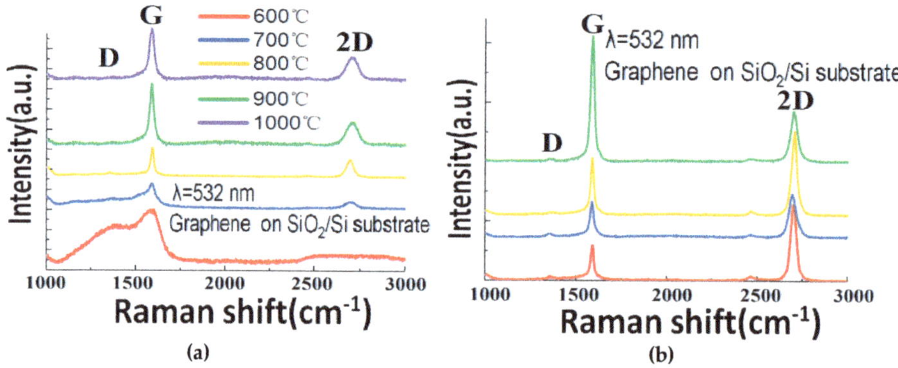


**Figure 4.** (**a**) Raman spectra of the graphene grown at different temperatures on Cu-Ni alloy and transferred to $SiO_2$/Si substrates. (**b**) Raman spectra captured at different positions in the same graphene sheet grown on a Cu-Ni alloy at 900 °C and transferred to its $SiO_2$/Si substrate.

In order to examine the structural uniformity of the graphene along the sample, more detailed characterizations are carried out. Figure 4b shows the Raman spectra measured at different positions of the same graphene sheet grown at 900 °C on a Cu-Ni alloy and transferred to its $SiO_2$/Si substrates. It is well established that the number of layers in graphene with good crystallinity can be estimated through the 2D/G ratio, and the shape and width of the peaks [25]. It can be seen that the number of graphene layers grown by the Cu-Ni alloy varies between monolayer, bilayer, and multilayers at different locations across the surface of the substrate. This is explained by the carbon segregation mechanism during the graphene growth on the metal [26], because the nickel content in the alloy has a high carbon solubility of 1.26 at.% [27] (see Figure 3a). Apart from the surface catalysis process, the dissolved carbon species can emit to the surface of the alloy upon cooling down, as a result of the reduced carbon solubility at lower temperature. Subsequently, at some areas the graphene layers are thicker due to the carbon segregation. If Cu is used as the catalyst, then the dominant growth mechanism is a surface catalytic graphitization procedure, because the carbon solubility is very low (only 0.0027 at.% [27]). Therefore, we have also grown graphene on copper foils without any Ni content, in order to obtain a large monolayer ratio, as will be shown thereinafter.

Figure 5a shows the Raman spectra of graphene samples (transferred to $SiO_2$/Si) grown on a 300 nm Cu-Ni alloy with different gas ratios at 900 °C. The gas mixture is varied from pure Ar, Ar:$H_2$ = 5:1, Ar:$H_2$ = 5:3, Ar:$H_2$ = 5:5, to pure $H_2$. It can be seen that the graphene quality almost does not change, which proves that the GD mechanism does not depend much on the Ar-$H_2$ composition of the gas. No matter whether it is a large atom Ar gas or a small molecule $H_2$ gas, the GD can etch the graphite electrode and supply carbon source anyway (although the etching mechanisms are different). Therefore, in future experiments, cheaper gas can be selected to prepare the graphene by this method. However, because the reaction temperature is typically above 800 °C, oxygen or air cannot be selected as the plasma gas so as to avoid the burning of the graphite.

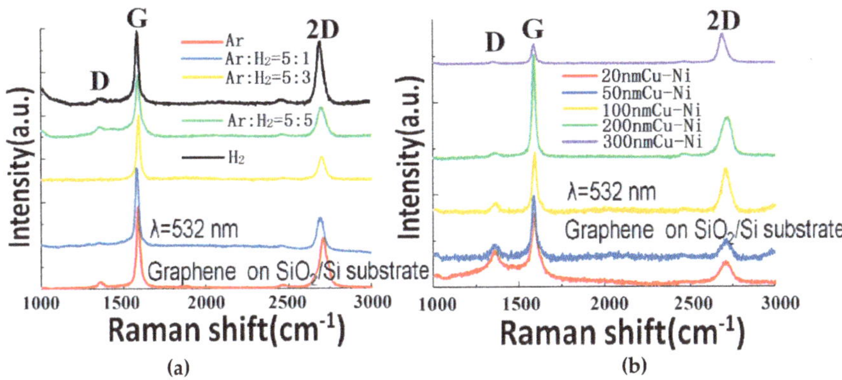


**Figure 5.** (**a**) Raman spectra of the graphene thin films grown with different gas ratios at 900 °C on 300 nm Cu-Ni and transferred to $SiO_2$/Si substrates. (**b**) Raman spectra of the graphene grown at 900 °C (transferred to $SiO_2$/Si) and Ar:$H_2$ (5:1) gas ratio with different thicknesses (from 20 to 300 nm) of Cu-Ni alloy

Figure 5b shows the Raman spectra of the graphene (transferred to $SiO_2$/Si) grown at 900 °C with a different thicknesses (from 20 to 300 nm) of Cu-Ni alloy and an Ar:$H_2$ = 5:1 gas ratio. The original idea of this experiment was to reduce the amount of absorbed carbon (hence reducing the number of produced graphene layers) through reducing the thickness of the Cu-Ni alloy. However, the experimental results show that even if the 20 nm Cu-Ni alloy is used, it still can produce bilayer and multilayer graphene. On the other hand, as the thickness decreases, the D peak rises and more defects appear. This article considers that the graphene quality gets worse at reducing thickness of the alloy because during the growth, the metals of the 20 nm and 50 nm samples sublimate from the surface under high temperature and low pressure conditions, which gradually leads to the loss of the graphene catalysis effect. If we continue to test the growth on even thinner metals, the graphene quality will be even worse, and the alloy will melt due the lowering of the melting point with the reduction of the thickness. Towards the other end, when the thickness of the Cu-Ni alloy reaches 100 nm and above, the quality of graphene has gradually stabilized.

As indicated earlier, we have also grown graphene on copper foils by the GD method in order to boost the monolayer ratio. The corresponding Raman measurement results are demonstrated in Figure 6. From Figure 6a, we can see that, by using the GD method, we can grow high-quality single-layer graphene on copper foil at 900 °C. When the growth temperature is increased to 1000 °C, however, some low quality and—most likely—bilayer graphene is seen to grow, and where the 2D/G ratio decreases, the D peak rises and the number of defects increases. The reason is not yet clear, but this paper believes it could be explained as follows. Compared with the copper foil at 900 °C, the surface of the copper foil at 1000 °C becomes rougher (see Figure 7) because it is closer to Cu's melting point 1083 °C. Eventually, the graphene quality deteriorates and some bilayer graphene starts to show up. The graphene quality is also worse compared with the graphene grown by the GD method on Cu-Ni at the same temperature (see Figure 5a). The effect is attributed to the fact that in graphene growth Ni has a much higher catalytic ability than Cu [28]. Figure 6b–d shows the Raman mapping data measured in a 28 μm × 28 μm (10 × 10 points) graphene area grown on copper foil at 900 °C and transferred to its $SiO_2$/Si substrate. The ratio of $I_D$ to $I_G$ is less than 0.35, and the ratio of $I_G/I_{2D}$ is mostly around 0.5. This proves that the uniformity of the number of graphene layers is very good across the sample and the graphene is mainly single layer. Figure 8 shows the optical transmittance of the graphene grown at 900 °C and then transferred to the glass substrate. It has a transmission rate of about 97.7% in almost the full wavelength band, in agreement with the expected value for standard monolayer graphene. Those results confirm that the GD technique proposed in this paper performs excellently during the graphene growth catalysis on copper foils at 900 °C.

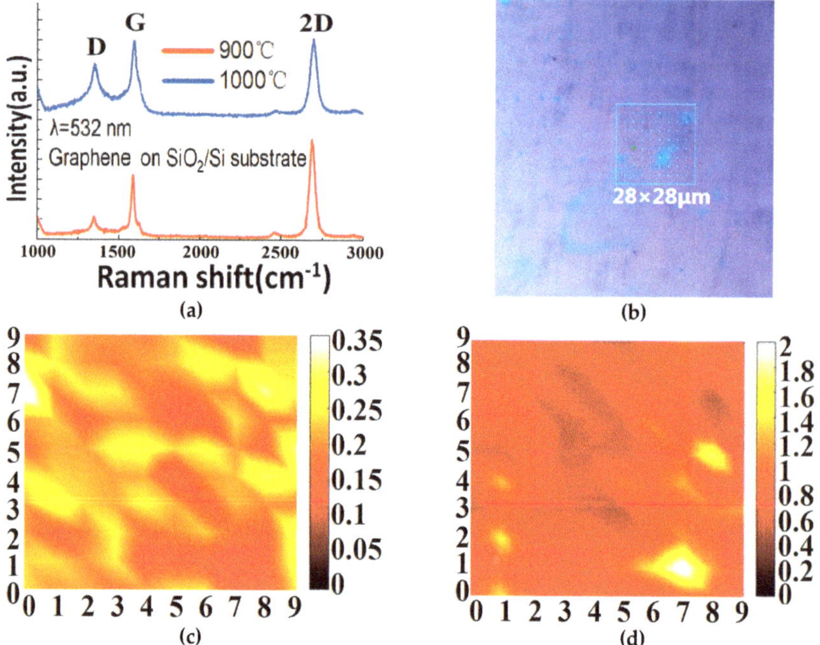


**Figure 6.** (**a**) Raman spectra of the graphene prepared by the GD method at 900 °C and 1000 °C by using copper foil as a catalyst. (**b**) Optical image (taken by Raman microscope) of a graphene thin film grown on copper foil at 900 °C and transferred to its $SiO_2$/Si substrate. The 28 μm × 28 μm part indicated by the square is the area for Raman mapping. (**c**,**d**) Raman mapping (28 μm × 28 μm) of the D/G and G/2D ratios of the graphene grown on copper foil at 900 °C and transferred to its $SiO_2$/Si substrate. In each image, there are 10 × 10 measured points and the color bar indicates the ratio.

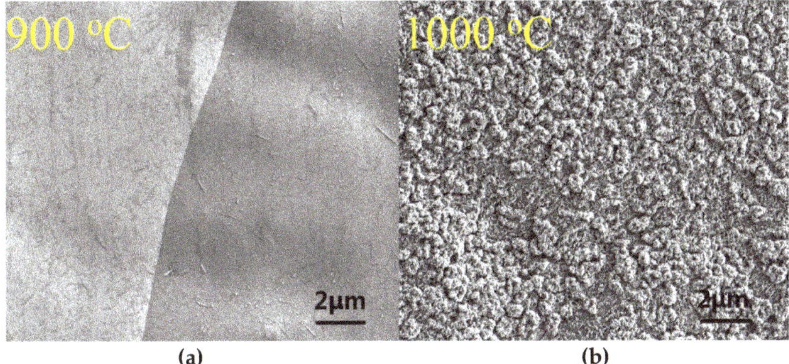


**Figure 7.** SEM images of the copper foils after coating with graphene at (**a**) 900 °C and (**b**) 1000 °C.

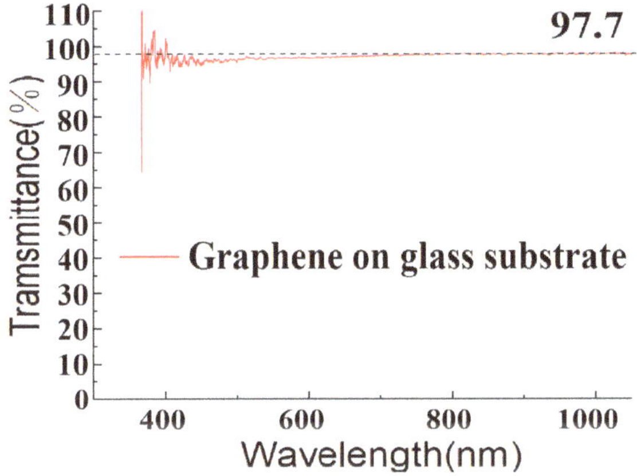


**Figure 8.** Optical transmittance of the as-grown graphene grown at 900 °C on Cu foil and transferred onto a glass substrate.

## 4. Conclusions

In this paper, we report a new type of graphene synthesis technology called GD deposition, using a standard PECVD machine. Compared to the traditional AD method, the GD technique uses a much smaller current, and the etching process of the graphite electrode that is used to ignite plasma is very gentle. The graphite electrode can be directly used as the carbon source for the graphene growth in a $H_2$ and/or Ar plasma environment, and high-quality graphene thin films can be synthesized at about 900 °C on catalytic metals e.g., Cu-Ni alloy and pure Cu. The graphene production mechanism is a plasma etching of the graphite, followed by a carbon containing particle emission, where the particles are scattered towards the catalytic metal surfaces for graphene thin film formation. To our knowledge, this graphene production mode has never been reported before. It reveals that graphene sheets (not irregular flakes) can be obtained by gaseous discharge-based technology. The as-grown graphene is transferrable, and is characterized by Raman spectroscopy, SEM, and other means to prove the feasibility of the GD. It is found that the major factors during the graphene synthesis that can determine the material quality are the growth temperature and the type of catalytic metals, but not the gas ratios. This work shows that the GD graphene production model can be used to prepare graphene thin films without using carbon-containing gases such as methane and ethylene, which provides a new technology and a new insight for graphene synthesis. In future work, we will continue to improve the quality of graphene and increase the sample size. We will apply the graphene in semiconductor devices, such as current spreading layers of light-emitting diodes [29]. We will also adjust the experimental conditions and explore the interesting topic, whether the GD method can produce vertical graphene [30] or carbon nanotubes.

**Author Contributions:** Conceptualization, L.W., Y.D. and J.S.; methodology, L.W., Y.D. and J.S.; formal analysis, W.G.; resources, Y.X., J.D. and C.X.; data curation, L.W., F.X., Z.D. and L.L.; writing—original draft preparation, L.W. and W.G.; writing—review and editing, J.S.; supervision, J.S.; project administration, J.S.; funding acquisition, W.G. and J.S. All authors have read and agreed to the published version of the manuscript.

**Funding:** This research was funded by the National Key R & D Program of China (2017YFB0403102) and the National Natural Science Foundation of China (11674016).

**Conflicts of Interest:** The authors declare no conflict of interest.

## References

1. Rao, C.N.R.; Subrahmanyam, K.S.; Matte, H.S.S.R.; Abdulhakeem, B.; Govindaraj, A.; Das, B.; Kumar, P.; Ghosh, A.; Late, D.J. A study of the synthetic methods and properties of graphenes. *Sci. Technol. Adv. Mater.* **2010**, *11*, 054502. [CrossRef] [PubMed]
2. Tzalenchuk, A.; Lara-Avila, S.; Kalaboukhov, A. Towards a quantum resistance standard based on epitaxial graphene. *Nat. Nanotechnol.* **2010**, *5*, 186–189. [CrossRef] [PubMed]
3. Cheng, T.S.; Davies, A.; Summerfield, A.; Cho, Y.; Cebula, I.; Hill, R.J.A.; Mellor, C.J.; Khlobystov, A.N.; Taniguchi, T.; Watanabe, K.; et al. High temperature MBE of graphene on sapphire and hexagonal boron nitride flakes on sapphire. *J. Vac. Sci. Technol. B* **2016**, *34*, 02L101. [CrossRef]
4. Sun, J.; Cole, M.T.; Lindvall, N.; Teo, K.B.K.; Yurgens, A. Noncatalytic chemical vapour deposition of graphene on high-temperature substrates for transparent electrodes. *Appl. Phys. Lett.* **2012**, *100*, 022102.
5. Wang, Z.; Li, N.; Shi, Z.; Gu, Z. Low-cost and large-scale synthesis of graphene nanosheets by arc discharge in air. *Nanotechnology* **2010**, *21*, 175602. [CrossRef]
6. KraTschmer, W.; Lamb, L.D.; Fostiropoulos, K.; Huffman, D.R. Solid $C_{60}$: A new form of carbon. *Nature* **1990**, *347*, 354–358. [CrossRef]
7. Pham, T.V.; Kim, J.; Jung, J.Y.; Kim, J.H.; Cho, H.; Seo, T.H.; Lee, H.; Kim, N.D.; Kim, M.J. High areal capacitance of N-Doped graphene synthesized by arc discharge. *Adv. Funct. Mater.* **2019**, *29*, 1905511. [CrossRef]
8. Volotskova, O.; Levchenko, I.; Shashurin, A.; Raitses, Y.; Ostrikov, K.; Keidar, M. Single-step synthesis and magnetic separation of graphene and carbon nanotubes in arc discharge plasmas. *Nanoscale* **2010**, *2*, 2281–2285. [CrossRef]
9. Huang, L.; Wu, B.; Chen, J.; Xue, Y.; Geng, D.; Guo, Y.; Yu, G.; Liu, Y. Gram-scale synthesis of graphene sheets by a catalytic arc-discharge method. *Small* **2013**, *9*, 1330–1335. [CrossRef]
10. Shashurin, A.; Keidar, M. Synthesis of 2D materials in arc plasmas. *J. Phys. D Appl. Phys.* **2015**, *48*, 314007. [CrossRef]
11. Ahmadi, R.; Ahmadi, M.T.; Ismail, R. Carbon nano-particle synthesized by pulsed arc discharge method as a light emitting device. *J. Electron. Mater.* **2018**, *47*, 4003–4009. [CrossRef]
12. Ismail, R.A.; Mohammed, M.I.; Mahmood, L.H. Preparation of multi-walled carbon nanotubes/n-Si heterojunction photodetector by arc discharge technique. *Opt. Int. J. Light Electron. Opt.* **2018**, *164*, 395–401. [CrossRef]
13. Wu, Z.S.; Ren, W.; Gao, L.; Zhao, J.; Chen, Z.; Liu, B.; Tang, D.; Yu, B.; Jiang, C.; Cheng, H.M. Synthesis of graphene sheets with high electrical conductivity and good thermal stability by hydrogen arc discharge exfoliation. *ACS Nano* **2009**, *3*, 411–417. [CrossRef] [PubMed]
14. Sun, J.; Lindvall, N.; Cole, M.; Angel, K.T.T.; Wang, T.; Teo, K.B.K.; Chua, D.H.C.; Liu, J.; Yurgens, A. Low partial pressure chemical vapour deposition of graphene on copper. *IEEE Trans. Nanotechnol.* **2012**, *11*, 255–260. [CrossRef]
15. De la Rosa, C.J.L.; Sun, J.; Lindvall, N.; Cole, M.T.; Nam, Y.; Löffler, M.; Olsson, E.; Teo, K.B.K.; Yurgens, A. Frame assisted $H_2O$ electrolysis induced $H_2$ bubbling transfer of large area graphene grown by chemical vapour deposition on Cu. *Appl. Phys. Lett.* **2013**, *102*, 022101. [CrossRef]
16. Takesaki, Y.; Kawahara, K.; Hibino, H.; Okada, S.; Tsuji, M.; Ago, H. Highly uniform bilayer graphene on epitaxial Cu-Ni(111) alloy. *Chem. Mater.* **2016**, *28*, 4583–4592. [CrossRef]

17. Dong, Y.; Xie, Y.; Xu, C.; Fu, Y.; Fan, X.; Li, X.; Wang, L.; Xiong, F.; Guo, W.; Pan, G.; et al. Transfer-free, lithography-free and fast growth of patterned CVD graphene directly on insulators by using sacrificial metal catalyst. *Nanotechnology* **2018**, *29*, 365301. [CrossRef]
18. Ishigami, M.; Chen, J.H.; Cullen, W.G.; Fuhrer, M.S.; Williams, E.D. Atomic structure of graphene on $SiO_2$. *Nano Lett.* **2007**, *7*, 1643–1648. [CrossRef]
19. Ago, H.; Ito, Y.; Mizuta, N.; Yoshida, K.; Hu, B.; Orofeo, C.M.; Tsuji, M.; Ikeda, K.; Mizuno, S. Epitaxial chemical vapour deposition growth of single-Layer graphene over cobalt film crystallized on sapphire. *ACS Nano* **2010**, *4*, 7407–7414. [CrossRef]
20. Bruce, C.E.R. Transition from glow to arc discharge. *Nature* **1948**, *161*, 521–522. [CrossRef]
21. Xie, L.; Jiao, L.; Dai, H. Selective etching of graphene edges by hydrogen plasma. *J. Am. Chem. Soc.* **2010**, *132*, 14751–14753. [CrossRef] [PubMed]
22. Davydova, A.; Despiau-Pujo, E.; Cunge, G.; Graves, D.B. Etching mechanisms of graphene nanoribbons in downstream $H_2$ plasmas: Insights from molecular dynamics simulations. *J. Phys. D Appl. Phys.* **2015**, *48*, 195202. [CrossRef]
23. Glad, X.; de Poucques, L.; Jaszczak, J.A.; Belmahi, M.; Ghanbaja, J.; Bougdira, J. Plasma synthesis of hexagonal-pyramidal graphite hillocks. *Carbon* **2014**, *76*, 330–340. [CrossRef]
24. Wu, S.; Yang, R.; Shi, D.; Zhang, G. Identification of structural defects in graphitic materials by gas-phase anisotropic etching. *Nanoscale* **2012**, *4*, 2005–2009. [CrossRef]
25. Ferrari, A.C.; Meyer, J.C.; Scardaci, V.; Casiraghi, C.; Lazzeri, M.; Mauri, F.; Piscanec, S.; Jiang, D.; Novoselov, K.S.; Roth, S.; et al. Raman spectrum of graphene and graphene layers. *Phys. Rev. Lett.* **2006**, *97*, 187401. [CrossRef]
26. Yu, Q.; Lian, J.; Siriponglert, S.; Li, H.; Chen, Y.P.; Pei, S. Graphene segregated on Ni surfaces and transferred to insulators. *Appl. Phys. Lett.* **2008**, *93*, 113103. [CrossRef]
27. Sun, J.; Nam, Y.; Lindvall, N.; Cole, M.; Kenneth, K.B.; Park, Y.W.; Yurgens, A. Growth mechanism of graphene on platinum: Surface catalysis and carbon segregation. *Appl. Phys. Lett.* **2014**, *104*, 152107. [CrossRef]
28. Dong, Y.; Guo, S.; Mao, H.; Xu, C.; Xie, Y.; Deng, J.; Wang, L.; Du, Z.; Xiong, F.; Sun, J. In situ growth of CVD graphene directly on dielectric surface toward application. *ACS Appl. Electron. Mater.* **2020**, *2*, 238–246. [CrossRef]
29. Xiong, F.; Guo, W.; Feng, S.; Li, X.; Du, Z.; Wang, L.; Deng, J.; Sun, J. Transfer-free graphene-like thin films on GaN LED epiwafers grown by PECVD using an ultrathin Pt catalyst for transparent electrode applications. *Materials* **2019**, *12*, 3533. [CrossRef]
30. Li, L.; Dong, Y.; Guo, W.; Qian, F.; Xiong, F.; Fu, Y.; Du, Z.; Xu, C.; Sun, J. High-responsivity photodetectors made of graphene nanowalls grown on Si. *Appl. Phys. Lett.* **2019**, *115*, 081101. [CrossRef]

*Article*

# Monolithic Integrated Device of GaN Micro-LED with Graphene Transparent Electrode and Graphene Active-Matrix Driving Transistor

**Yafei Fu [1], Jie Sun [2,*], Zaifa Du [1], Weiling Guo [1,*], Chunli Yan [3], Fangzhu Xiong [1], Le Wang [1], Yibo Dong [1], Chen Xu [1], Jun Deng [1], Tailiang Guo [2] and Qun Yan [2]**

1 Key Laboratory of Optoelectronics Technology, College of Microelectronics, Beijing University of Technology, Beijing 100124, China; fuxixi321@163.com (Y.F.); 17801011216@163.com (Z.D.); fangzhuxiong@emails.bjut.edu.cn (F.X.); wangle316@emails.bjut.edu.cn (L.W.); donyibo@emails.bjut.edu.cn (Y.D.); xuchen58@bjut.edu.cn (C.X.); dengsu@bjut.edu.cn (J.D.)

2 National and Local United Engineering Laboratory of Flat Panel Display Technology, College of Physics and Information Engineering, Fuzhou University, Fuzhou 350116, China; gtl@fzu.edu.cn (T.G.); qunfyan@gmail.com (Q.Y.)

3 Department of Information and Automation, Library of Fuzhou University, Fuzhou 350116, China; yan_chunli@yahoo.com

* Correspondence: jie.sun@fzu.edu.cn (J.S.); guoweiling@bjut.edu.cn (W.G.)

Received: 31 December 2018; Accepted: 22 January 2019; Published: 30 January 2019

**Abstract:** Micro-light-emitting diodes (micro-LEDs) are the key to next-generation display technology. However, since the driving circuits are typically composed of Si devices, numerous micro-LED pixels must be transferred from their GaN substrate to bond with the Si field-effect transistors (FETs). This process is called massive transfer, which is arguably the largest obstacle preventing the commercialization of micro-LEDs. We combined GaN devices with emerging graphene transistors and for the first-time designed, fabricated, and measured a monolithic integrated device composed of a GaN micro-LED and a graphene FET connected in series. The *p*-electrode of the micro-LED was connected to the source of the driving transistor. The FET was used to tune the work current in the micro-LED. Meanwhile, the transparent electrode of the micro-LED was also made of graphene. The operation of the device was demonstrated in room temperature conditions. This research opens the gateway to a new field where other two-dimensional (2D) materials can be used as FET channel materials to further improve transfer properties. The 2D materials can in principle be grown directly onto GaN, which is reproducible and scalable. Also, considering the outstanding properties and versatility of 2D materials, it is possible to envision fully transparent micro-LED displays with transfer-free active matrices (AM), alongside an efficient thermal management solution.

**Keywords:** GaN micro-light-emitting diodes; two-dimensional materials; graphene; field effect transistors; monolithic integration

---

## 1. Introduction

With the rapid development of modern intelligent devices, the demand for power reduction and resolution enhancement of electronic displays has become higher and higher. Light-emitting diodes, or micro-LEDs, have a single-pixel area $\leq 2.5 \times 10^{-3}$ mm$^2$ (50 $\times$ 50 µm or less). They represent the development trend of the next generation of displays by virtue of their low power, weather fastness, and super-high resolution. Currently, micro-LEDs constitute a new research focus in major university labs and in the semiconductor industry. Nevertheless, as it is still being developed, the technology has not yet been commercialized. One of the major bottlenecks is the "massive transfer" [1]. Micro-LEDs are typically fabricated on gallium nitride (GaN)-based wafers, whereas the active matrix (AM) circuitry is made of Si-based complementary metal-oxide-semiconductor (CMOS) devices. A massive transfer of hundreds of thousands of micro-LED pixels is needed to bond the micro-LEDs with their individual AM driving elements and this is a huge technical obstacle, since the process relies critically on alignment accuracy, bonding strength, and reliability. It is expected that the alignment accuracy should be within $\pm$ 0.5 μm, and the yield higher than 99.9999%. Currently, most researchers are focusing on developing advanced transfer technologies. However, we believe that another route that is very promising and should receive more attention is the monolithic integration of micro-LEDs and their driving circuits. This route can bypass the massive transfer, reduce cost and technological difficulties, and be more reliable in applications. Previously, GaN-based micro LEDs were integrated with GaN high-electron-mobility transistors (HEMTs) [2] or metal-oxide-semiconductor field-effect transistors (MOSFETs) [3]. GaN transistors generally have high breakdown voltages [4,5] and high working frequencies [6,7]. However, because of the lattice mismatch, it is hard to grow high-quality GaN LEDs and transistors directly on top of each other. Furthermore, the deposition temperatures of the two are quite different, making the growth processes of the two materials not compatible [2,3].

Graphene, a monolayer of graphite, is known as a new wonder material. It integrates many outstanding properties into one material (e.g., high carrier mobility, minimal thickness, high optical transmittance, high thermal conductivity, excellent chemical stability, mechanical strength, and flexibility) and hence, has a great potential in nanoelectronics [8]. It can be used as the channel material for transparent high-frequency transistors [9]. Furthermore, it can be used in transparent electrodes for LEDs [10]. Compared with traditional indium tin oxide (ITO), graphene electrodes have a broader spectrum and are more mechanically flexible and chemically stable. Graphene has been used in GaN LEDs and has shown potential for solving the current crowding problem caused by the resistive *p*-GaN capping layer of LEDs [11,12].

In this paper, we have designed and fabricated GaN micro-LEDs with graphene transparent electrodes and graphene field-effect driving transistors (GFET). To the best of our knowledge, this is the first proof-of-principle study on the monolithic integration of GaN micro-LEDs with their AM driving GFET. The role of graphene is two-fold: it acts as a transparent conducting film for the micro-LEDs and also as the channel material for the driving FETs to control the micro-LEDs. We demonstrate that graphene grown by chemical vapor deposition (CVD) can effectively spread the current in the *p*-GaN, making the lighting more uniform, while the field effect in the graphene can be used to tune the current in the integrated device. This technique is intrinsically scalable and compatible with the semiconductor industry. Although the present study uses graphene grown ex situ on a Cu foil, the method can be extended to graphene grown in situ on GaN [13]. Also, channel material may not be limited to graphene. Considering recent advances in large-scale growth of other two-dimensional (2D) materials, e.g., $MoS_2$ with higher transistor on–off ratios [14], this research can be viewed as pioneering work which reveals the promising future of combining emerging 2D materials with traditional bulk semiconductors to resolve the issue of massive transfer that exists currently in the micro-LED community.

## 2. Experimental Procedures

Figure 1 is a schematic illustration of the structure of the integrated GaN micro-LED and GFET device, which was fabricated on a commercial GaN LED epitaxial wafer with c-face-patterned sapphire substrate (PSS). The epi-layers were grown by metalorganic chemical vapor deposition (MOCVD) using a Veeco K465i system. From bottom to top, on the 17 nm buffer layer, there were in turn 2.5 μm *n*-GaN, 74 nm *n*-AlGaN, 131.75 nm multiple quantum well (MQW, InGaN/GaN), 40.62 nm electron-blocking layer (six periods of AlGaN/GaN = 3.61 nm/3.16 nm superlattice), and 92 nm *p*-GaN. The device fabrication process flow was as follows. First, mesa areas of the GaN micro-LEDs were defined by lift-off photolithography carried out in 2 nm Pt with 110 nm Ni sputtered onto the GaN LED wafer. The Pt/Ni layer was on the GaN mesa top, where Pt was used to improve the electrical contact between the transparent electrode and *p*-GaN, and Ni was used as the hard mask in the dry-etching.

Inductively coupled plasma (ICP) etching was used to etch the semiconductor down to the *n*-GaN (1.25 μm deep) using a $SiH_4/Cl_2$ gas mixture. The Ni mask was then wet-etched away. At 300 °C, plasma-enhanced CVD (PECVD) was used to grow 300 nm $SiO_2$ onto the sample as an insulation layer between the micro-LED and the graphene transistor. The patterning of this $SiO_2$ layer was achieved by using buffered oxide etch (BOE). The buried back gate of the transistor was fabricated by another lift-off lithography with a sputtered 15 nm Ti/100 nm Au metal layer, as shown in Figure 1. PECVD was used for a second time to grow 300 nm $SiO_2$, followed by BOE patterning to bury the gate. This layer of silicon dioxide served as the gate dielectric. Afterwards, a third lift-off lithography was used to pattern the sputtered 15 nm Ti/130 nm Au in order to form the *p*, *n* metal electrodes of the micro-LED, as well as the source and drain of the GFET. The *p*-pad of the micro-LED was connected to the source of the transistor. Finally, the graphene was grown by standard CVD on Cu foil using $CH_4$ as the precursor (Shenzhen Jingge Nano Technology, Shenzhen, China). After the graphene CVD, the Cu foil was etched off, and the graphene transferred to the GaN wafer using standard graphene transfer technology [15]. The graphene, patterned by photolithography and oxygen plasma etching, functioned as the transparent electrode on the micro-LED and the channel in the driving FET.

The inset of Figure 1 is the equivalent circuit of the integrated micro-LED/GFET device. The micro-LED was essentially a GaN *pn* junction inserted with multiple quantum wells. When the injected current was large enough, spontaneous emission of photons was initiated by the recombination of electron–hole pairs. The GFET was connected with the micro-LED in series. Its gate capacitively coupled to the graphene channel. The Fermi energy in the channel was tuned by controlling the gate voltage, which was then translated into the modulation of the channel current $I_d$. In this way, the current flowing into the micro LED and thus the electroluminescence could be adjusted.

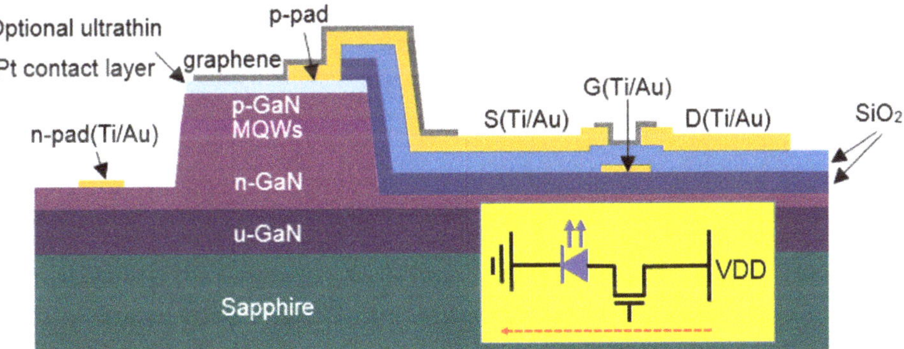


**Figure 1.** Schematic diagram of the monolithic integrated micro-light-emitting diodes (micro-LED)/graphene field-effect driving transistors (GFET) device. Some semiconductor layers are omitted for simplicity. The inset is its equivalent circuit, where the micro-LED is connected in series with the driving transistor (AM). The red arrow indicates the direction of the current flow. This figure is not drawn to scale. GaN, $V_{DD}$, G, D, S and MQW denote gallium nitride, total applied voltage of the integrated device, gate, drain, source and multiple quantum well, respectively.

## 3. Results and Discussion

Figure 2a shows an optical micrograph of the micro-LED/GFET device. The mesa area of the micro LED was 30 μm × 50 μm, and the graphene channel of the driving transistor was 12 μm × 480 μm. The *p* electrode of the micro-LED was connected to the source pad of the transistor. The S, D metal pads of the GFET were sitting on top of the double-layer $SiO_2$ film. It can be seen that the G pad was a little blurred, because the gate was buried in between the two $SiO_2$ layers, as shown in Figure 1. Figure 2b is a typical Raman spectrum measured in graphene. The G band and 2D band peaks at ~1600 $cm^{-1}$ and ~2690 $cm^{-1}$ are signatures of $sp^2$ hybridized graphitic carbon. The $I_G/I_{2D}$ ratio was approximately

1/2, and the full width at half maximum (FWHM) of the 2D peak was ~37.92 $cm^{-1}$, indicating the graphene was a monolayer [16]. The D peak is exceedingly small, meaning that there were few disorders in the graphene lattice; therefore, the quality of the as-grown graphene was satisfactory.

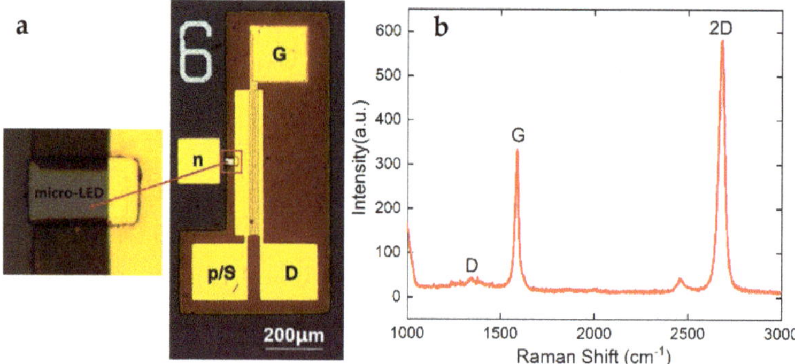


**Figure 2.** (**a**) Optical microscopy image of the micro-LED/GFET-integrated device. The mesa of the GaN micro LED is magnified; (**b**) Raman spectrum of the graphene monolayer.

First, we measured discrete components' performances. The static (DC) properties of the GFET were characterized at room temperature without special treatment, e.g., vacuum annealing. Figure 3a,b shows the output and transfer properties, respectively. In the GFET, 0–4 V ($V_d$) was swept between the source and drain, while the gate voltage $V_g$ was altered in steps (−40 V, 0 V, 20 V, and 40 V). In Figure 3a, one can see that with increasing gate voltage, the slope of the output curves got smaller, indicating an increasing channel resistance. Clearly, the channel was *p*-type. When $V_d$ was fixed at 0.1 V and the gate voltage $V_g$ was swept between −40 V and 40 V, the transfer curve shown in Figure 3b could be obtained. The Dirac point was at $V_g > 40$ V, which means the graphene was far from intrinsic (heavy *p*-doping). This can primarily be ascribed to the photolithography procedure, where the photoresist residue is known to dope the graphene severely [17]. Unfortunately, this doping significantly damages graphene mobility. Apart from the doping effect introduced during the device processing, direct exposure to open air without passivation leads to adsorption of $H_2O$, which also reduces the carrier mobility. Using a formula based on a capacitor model:

$$\mu = \frac{\delta I_d}{C_g V_d \delta V_g} \frac{L}{W} \tag{1}$$

where $\mu$, $C_g$, $L$, $W$ are the carrier mobility, gate capacitance, gate length, and gate width, respectively, the hole mobility in the GFET was calculated to be 696 $cm^2 \cdot V^{-1} \cdot s^{-1}$ (the value $\frac{\delta I_d}{\delta V_g}$ was taken when $V_g$ = 25 V). We note that this is the field-effect mobility, and the calculation depends on many device parameters, e.g., interface trap density. Furthermore, we used a simplified model without considering the quantum capacitance. The standard Hall measurement gave a more accurate estimation of graphene mobility on polyethylene terephthalate (PET) substrate, showing 1042 $cm^2 \cdot V^{-1} \cdot s^{-1}$. The transconductance $g_m$ of the GFET could be directly calculated from Figure 3b. At $V_g$ = 25 V, the transconductance reached its maximum, which was about 0.067 mS/mm (normalized to the gate width). The sheet resistance of the monolayer CVD graphene was approximately 4 kΩ/□.

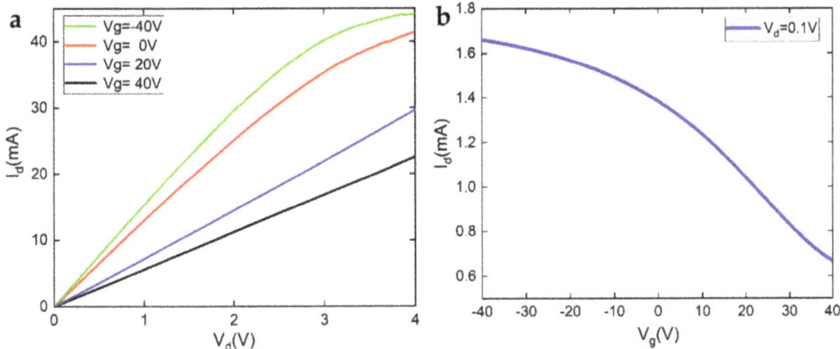


**Figure 3.** (**a**) Output and (**b**) transfer properties of the GFET measured at room temperature.

Graphene is a semimetal with no bandgap. Thus, graphene–metal contacts are generally very good. The specific contact resistivity for typical graphene–metal contacts is $10^{-5}$ or $10^{-6}$ $\Omega cm^2$ [18], with some of the best examples being $10^{-7}$ $\Omega cm^2$ or lower [19]. For our graphene, the typical value is $\varrho_c = 1 \times 10^{-6}$ $\Omega cm^2$ [20]. The total overlapping area in the integrated device was 480 μm × 55 μm, which means the total contact resistances arising from the graphene–metal contacts added up to less than 0.01 Ω. Clearly, this figure can be ignored. Good graphene–metal contact behaviour was also confirmed by another work of ours [11].

Figure 4 plots the I–V properties of the GaN micro-LED. The black and red curves are the logarithmic and linear plots, respectively. In this figure, the turn-on voltage is shown to be around 5.8 V, a higher voltage than that of commercial GaN LED devices. This was of course due to our fabrication technique but, more importantly, to the fact that the graphene's Fermi level did not match that of *p*-GaN. The forward differential resistance (series resistance) of the micro-LED was estimated to be as large as 190 Ω, which was also due to the nonoptimal graphene–GaN contact. The 2 nm Pt layer can bridge the Fermi levels of the graphene and *p*-GaN, improving the Ohmic contact to some extent. Currently, there is no stable and mature doping technology for graphene. In the future, if effective and stable doping can be realized, the graphene work function could be greatly improved, and the turn-on voltage would ultimately be reduced. At 8 V forward voltage $V_{LED}$, the work current $I_d$ was about 10.5 mA, which is large enough for a micro-LED. The insert of Figure 4 is a photograph taken when the micro-LED was turned on. It is an intuitive image demonstrating the uniformity and high intensity of the lighting. Because of the much higher conductivity of graphene compared with *p*-GaN, the graphene transparent electrode helped spread the current effectively along the *p*-GaN mesa surface, solving the current crowding problem. The high transmittance of graphene ensured the light was effectively emitted to the external world.

Finally, we examined the properties of the GaN micro-LED/GFET as an integrated device. $V_{DD}$ of 0–9 V was swept on the drain electrode of the GFET, with the *n* electrode of the micro-LED grounded, as shown in the inset of Figure 1. The I–V properties are plotted in Figure 5a, where the gate voltage $V_g$ is shown at −40 V, 0 V, 20 V, and 40 V. In this figure, while tuning $V_g$, the current $I_d$ showed some changes when the $V_{DD}$ was relatively large. The current reached its maximum when $V_g$ was −40 V and reached its minimum at $V_g$ = 40 V. When $V_{DD}$ was 9 V, the difference between $I_{max}$ and $I_{min}$ was approximately 3.5 mA. This confirmed that the GFET can indeed be used to control the micro-LED in an AM fashion. Figure 5b plots the I–V curves of the micro-LED and its graphene driving transistor in the same figure and can be used to determine the static working point (the crossover points in this figure) of the integrated device. In this figure, $V_{DD}$ was set to be 8 V,(maximum voltage on the horizontal axis). The device current $I_d$ (work current of both the micro-LED and the GFET) was plotted against the voltage on the transistor source, $V_s$, which is also the voltage on the *p* pad of the micro-LED. Clearly, when $V_g$ was tuned from negative to positive, the voltage of the crossover point, $V_s$, was reduced, and the current $I_d$ was reduced correspondingly. This is because the graphene channel resistance increases when the gate voltage becomes more positive, due to the hole conduction mechanism. The voltage drop across the GFET ($V_{DD}$–$V_s$) in the series circuit will therefore increase, leading to a reduction of voltage on the micro-LED. As the I–V relation of the micro-LED is exponential, a small decrease in the voltage on the $p$ electrode, $V_s$, will result in a big decrease in the work current $I_d$. This notwithstanding, compared to micro-LEDs driven by MOSFETs [3], the current tunability in our device was modest. At the gate voltage range shown in Figure 5, the micro-LED could not be totally turned off ($V_s$ cannot be < 5.8 V). This can be attributed to the process-deteriorated graphene quality (mainly mobility). Furthermore, graphene is gapless, which means the GFET has a low on–off ratio. Also, the buried gate was very far from the channel, and the gate dielectric was not high-k material (300 nm $SiO_2$). In the future, the channel material of the AM driving transistors can consist of other emerging 2D materials that have energy bandgaps, such as $MoS_2$ and other transition-metal dichalcogenides (TMDs). With the technology of 2D materials becoming more and more developed, mobility, and consequently the transconductance of the FET channel, can be much improved. The size and aspect ratio of the gate can be optimized, and the gate dielectric can be replaced by high-k insulators [21] to further boost transconductance.

The integrated devices reported in this paper and in reference [3] can both bypass the technical obstacles of massive transfer, accurate alignment, and bonding of micro-LED pixels. This bypass is a great advantage when compared with devices with a traditional MOSFET driving module fabricated on a separate Si wafer. When comparing our work with that described in reference [3], the performances of the micro-LEDs are very similar. The final integrated devices also work at similar voltage and current ranges. However, our GFET transconductance is four orders of magnitude lower than that of GaN MOSFETs [3], resulting in a much weaker tunability. On the other hand, unlike GaN MOSFETs, which have been researched for decades, 2D material FETs are still in their infancy, and there is still a lot of room to improve, as was discussed earlier. Most importantly, our device concept and fabrication technology are much simpler than those in reference [3], where two GaN dry-etching steps of different depths are required to fabricate the micro-LED and the MOSFET, respectively.

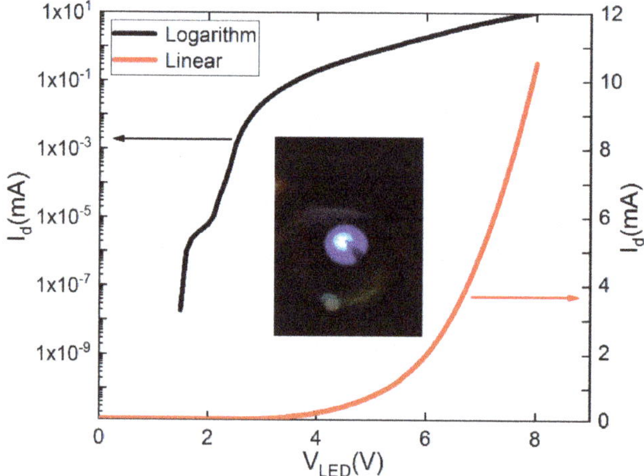


**Figure 4.** Current–voltage characteristics of the micro-LED plotted in linear and logarithmic scales. The inset is an electroluminescence photo of the device measured in a probe station.

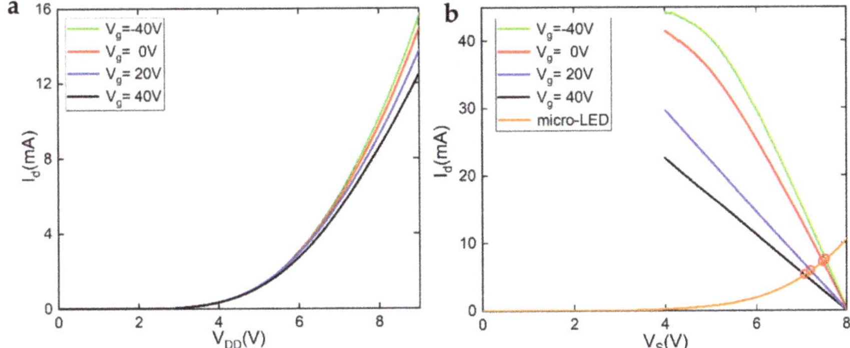


**Figure 5.** (**a**) The overall I–V curve of the integrated micro-LED/GFET device; (**b**) demonstration of the static working mechanism of the integrated device. The device current is plotted against $V_s$, with $V_{DD}$ fixed at 8 V. The crossing points are referred to as static work points in this paper.

## 4. Conclusions

In this paper, we have designed, fabricated, and characterized a monolithic integrated device composed of a GaN micro-LED and a driving element, namely, a graphene transistor. Meanwhile, the graphene on the GaN mesa served as the transparent electrode for the micro-LED. This is the first demonstration of combining GaN micro-LEDs with AM GFETs to bypass the technical difficulty of massive transfer, which exists currently in the micro-LED community. By tuning the gate voltage of the GFET, the micro-LED current and the light intensity could be adjusted. The performance of the integrated device needs further improvement. With research booming on 2D materials other than graphene, we believe this research opens the gateway for combining traditional bulk semiconductors with ultrathin 2D materials in a bid to improve micro-LED applications. The 2D materials can be grown in situ onto GaN to achieve better process reproducibility. Furthermore, since 2D materials such as graphene often have very high thermal conductivity, the heat spreading effect should not be overlooked. This is especially important for micro-LED displays, where the pixel density is super-high, and an appropriate thermal management solution is a must. The thermal aspect of the monolithic integrated device is a subject for future work. Additionally, as 2D materials are atomically thin and transparent, as long as the metal electrodes are replaced by transparent conductors, e.g., ITO, the whole display can be made transparent, which is well in line with the future trend of development.

**Author Contributions:** Data Curation, Z.D., F.X. and L.W.; Methodology, W.G. and J.D.; Supervision, J.S., C.X. and Q.Y.; Writing—Original Draft, Y.F. and J.S.; Writing—Review & Editing, C.Y., Y.D. and T.G.

**Funding:** This research was funded by National Natural Science Foundation of China, grant number 11674016; National key R & D program, grant number 2017YFB0403100, 2017YFB0403102 and 2018YFA0209000; Beijing Municipal Commission of Science and Technology, grant number Z1611 00002116032.

**Acknowledgments:** J.S. would like to thank Start Up Package of Fuzhou University.

**Conflicts of Interest:** The authors declare no conflict of interest.

## References

1. Lee, V.W.; Twu, N.; Kymissis, I. Micro-LED Technologies and Applications. *Inf. Disp.* **2016**, *6*, 16–23. [CrossRef]
2. Liu, Z.J.; Huang, T.; Ma, J.; Liu, C.; Lau, K.M. Monolithic Integration of AlGaN/GaN HEMT on LED by MOCVD. *IEEE Electron Device Lett.* **2014**, *35*, 330–332. [CrossRef]
3. Lee, Y.J.; Yang, Z.P.; Chen, P.G.; Hsieh, Y.A.; Yao, Y.C.; Liao, M.H.; Lee, M.H.; Wang, M.T.; Hwang, J.M. Monolithic integration of GaN-based light-emitting diodes and metal-oxide-semiconductor field-effect transistors. *Opt. Express* **2014**, *22* (Suppl. 6), A1589–A1595. [CrossRef]

4. Wu, Y.F.; Keller, B.P.; Keller, S.; Kapolnek, D.; Kozodoy, P.; Denbaars, S.P.; Mishra, U.K. Very high breakdown voltage and large transconductance realized on GaN heterojunction field effect transistors. *Appl. Phys. Lett.* **1996**, *69*, 1438–1440. [CrossRef]
5. Dora, Y.; Chakraborty, A.; McCarthy, L.; Keller, S.; Denbaars, S.P.; Mishra, U.K. High Breakdown Voltage Achieved on AlGaN/GaN HEMTs With Integrated Slant Field Plates. *IEEE Electron Device Lett.* **2006**, *27*, 713–715. [CrossRef]
6. Sheppard, S.T.; Doverspike, K.; Pribble, W.L.; Allen, S.T.; Palmour, J.W.; Kehias, L.T.; Jenkins, T.J. High-Power Microwave GaN/AlGaN HEMTs on Semi-Insulating Silicon Carbide Substrates. *IEEE Electron Device Lett.* **1999**, *20*, 161–163. [CrossRef]
7. Saadat, O.I.; Chung, J.W.; Piner, E.L.; Palacios, T. Gate-First AlGaN/GaN HEMT Technology for High-Frequency Applications. *IEEE Electron Device Lett.* **2009**, *30*, 1254–1256. [CrossRef]
8. Novoselov, K.S.; Fal'ko, V.I.; Colombo, L.; Gellert, P.R.; Schwab, M.G.; Kim, K. A roadmap for graphene. *Nature* **2012**, *490*, 192–200. [CrossRef]
9. Schwierz, F. Graphene transistors. *Nat. Nanotechnol.* **2010**, *5*, 487–496. [CrossRef]
10. Bonaccorso, F.; Sun, Z.; Hasan, T.; Ferrari, A.C. Graphene photonics and optoelectronics. *Nat. Photonics* **2010**, *4*, 611–622. [CrossRef]
11. Xu, K.; Xu, C.; Deng, J.; Zhu, Y.; Guo, W.; Mao, M.; Zheng, L.; Sun, J. Graphene transparent electrodes grown by rapid chemical vapor deposition with ultrathin indium tin oxide contact layers for GaN light emitting diodes. *Appl. Phys. Lett.* **2013**, *102*, 162102. [CrossRef]
12. Xu, K.; Xu, C.; Xie, Y.; Deng, J.; Zhu, Y.; Guo, W.; Mao, M.; Xun, M.; Chen, M.; Zheng, L.; et al. GaN nanorod light emitting diodes with suspended graphene transparent electrodes grown by rapid chemical vapor deposition. *Appl. Phys. Lett.* **2013**, *103*, 222105. [CrossRef]
13. Sun, J.; Cole, M.T.; Ahmad, S.A.; Backe, O.; Ive, T.; Loffler, M.; Lindvall, N.; Olsson, E.; Teo, K.B.K.; Liu, J.; et al. Direct Chemical Vapor Deposition of Large-Area Carbon Thin Films on Gallium Nitride for Transparent Electrodes: A First Attempt. *IEEE Trans. Semicond. Manuf.* **2012**, *25*, 494–501. [CrossRef]
14. Sun, J.; Li, X.; Guo, W.; Zhao, M.; Fan, X.; Dong, Y.; Xu, C.; Deng, J.; Fu, Y. Synthesis Methods of Two-Dimensional $MoS_2$: A Brief Review. *Crystals* **2017**, *7*, 198. [CrossRef]
15. Liu, L.; Liu, X.; Zhan, Z.; Guo, W.; Xu, C.; Deng, J.; Chakarov, D.; Hyldgaard, P.; Schröder, E.; Yurgens, A. A Mechanism for Highly Efficient Electrochemical Bubbling Delamination of CVD-Grown Graphene from Metal Substrates. *Adv. Mater. Interfaces* **2016**, *3*, 1500492. [CrossRef]
16. Ferrari, A.C.; Meyer, J.C.; Scardaci, V.; Casiraghi, C.; Lazzeri, M.; Mauri, F.; Piscanec, S.; Jiang, D.; Novoselov, K.S.; Roth, S.; et al. Raman Spectrum of Graphene and Graphene Layers. *Phys. Rev. Lett.* **2006**, *97*, 187401. [CrossRef] [PubMed]
17. Shi, R.; Xu, H.; Chen, B.; Zhang, Z.; Peng, L.M. Scalable fabrication of graphene devices through photolithography. *Appl. Phys. Lett.* **2013**, *102*, 113102. [CrossRef]
18. Robinson, J.; LaBella, M.; Zhu, M.; Hollander, M.; Kasarda, R.; Hughes, Z.; Trumbull, K.; Cavalero, R.; Snyder, D. Contacting graphene. *Appl. Phys. Lett.* **2011**, *98*, 053103. [CrossRef]
19. Moon, J.; Antcliffe, M.; Seo, H.; Curtis, D.; Lin, S.; Schmitz, A.; Milosavljevic, I.; Kiselev, A.; Ross, R.; Gaskill, D.; et al. Ultra-low resistance ohmic contacts in graphene field effect transistors. *Appl. Phys. Lett.* **2012**, *100*, 203512. [CrossRef]
20. Andersson, M.; Vorobiev, A.; Sun, J.; Yurgens, A.; Gevorgian, S.; Stake, J. Microwave characterization of Ti/Au-graphene contacts. *Appl. Phys. Lett.* **2013**, *103*, 173111. [CrossRef]
21. Sun, J.; Lind, E.; Maximov, I.; Xu, H.Q. Memristive and Memcapacitive Characteristics of a Au/Ti–$HfO_2$-InP/InGaAs Diode. *IEEE Electron Device Lett.* **2011**, *32*, 131–133. [CrossRef]

*Article*

# Transfer-Free Graphene-Like Thin Films on GaN LED Epiwafers Grown by PECVD Using an Ultrathin Pt Catalyst for Transparent Electrode Applications

**Fangzhu Xiong [1], Weiling Guo [1,*], Shiwei Feng [1], Xuan Li [1], Zaifa Du [1], Le Wang [1], Jun Deng [1] and Jie Sun [1,2,*]**

1 Key Laboratory of Optoelectronics Technology, College of Microelectronics, Beijing University of Technology, Beijing 100124, China; fangzhuxiong@emails.bjut.edu.cn (F.X.); shwfeng@bjut.edu.cn (S.F.); lixuan@emails.bjut.edu.cn (X.L.); 17801011216@163.com (Z.D.); wangle316@emails.bjut.edu.cn (L.W.); dengsu@bjut.edu.cn (J.D.)

2 Quantum Device Physics Laboratory, Department of Microtechnology and Nanoscience, Chalmers University of Technology, 41296 Gothenburg, Sweden

* Correspondence: guoweiling@bjut.edu.cn (W.G.); jie.sun@chalmers.se (J.S.)

Received: 25 September 2019; Accepted: 24 October 2019; Published: 28 October 2019

**Abstract:** In this work, we grew transfer-free graphene-like thin films (GLTFs) directly on gallium nitride (GaN)/sapphire light-emitting diode (LED) substrates. Their electrical, optical and thermal properties were studied for transparent electrode applications. Ultrathin platinum (2 nm) was used as the catalyst in the plasma-enhanced chemical vapor deposition (PECVD). The growth parameters were adjusted such that the high temperature exposure of GaN wafers was reduced to its minimum (deposition temperature as low as 600 °C) to ensure the intactness of GaN epilayers. In a comparison study of the Pt-GLTF GaN LED devices and Pt-only LED devices, the former was found to be superior in most aspects, including surface sheet resistance, power consumption, and temperature distribution, but not in optical transmission. This confirmed that the as-developed GLTF-based transparent electrodes had good current spreading, current injection and thermal spreading functionalities. Most importantly, the technique presented herein does not involve any material transfer, rendering a scalable, controllable, reproducible and semiconductor industry-compatible solution for transparent electrodes in GaN-based optoelectronic devices.

**Keywords:** transfer-free; PECVD; graphene; gallium nitride; LEDs; transparent electrodes; heat spreading

## 1. Introduction:

Gallium nitride (GaN) has attracted remarkable attention as an important material for application in optoelectronic and microelectronic devices, such as light-emitting diodes (LEDs), laser diodes (LDs), solar cells (SCs), and high electron mobility transistors (HEMTs) [1–4]. In order to simultaneously improve the current spreading, current injection and light extraction efficiency, transparent electrodes are commonly used in GaN-based LEDs and SCs. In current optoelectronic devices, the mainstream transparent electrode material is indium tin oxide (ITO) [5]. However, ITO has an increasingly high price as indium is slathered and getting scarce. Also, ITO is nontransparent at very low wavelength regimes and has poor chemical stability, which is not suitable for ultraviolet GaN LEDs [6,7]. Since 2004 [8], graphene has shown great potential in the fields of nanoelectronics, energy, chemistry and biomedicine for its excellent properties, such as high transparency, conductivity, mobility, thermal conductivity and mechanical strength. Graphene is wide-spectrum transparent (from ultraviolet to near infrared) [9] and for every additional layer, the transparency ideally only decreases by 2.3%.

In addition, graphene's preparation process is relatively simple and cheap. As a result, graphene is likely to be a substitute for ITO [10,11]. At present, copper is widely used to catalyze the growth of graphene by chemical vapor deposition (CVD) because copper has low carbon solubility and thus it is easy to form a single layer of graphene. Kim et al. [12] used Cu as a catalyst to grow graphene. At a 372 nm wavelength, two layers of graphene had a transmittance up to 95%, and four layers were up to 89%, better than 372 nm thick ITO's 68% transmittance at this wavelength range. However, this method requires a process of wet or dry transfer of graphene to the new substrate after growth, which is very complex and often leads to non-ideal interfaces between graphene and gallium nitride, such as metal residues, oxides, holes and wrinkles. Currently, few people have studied graphene on platinum [13–18]. According to Gao et al. [13], compared to copper, platinum has a stronger catalytic capacity on hydrocarbon decomposition and subsequent graphene formation. Therefore, in this study, we used ultrathin platinum (2 nm) as a catalyst for direct growth (i.e., transfer-free) of graphene-like thin films (GLTFs) on a GaN/sapphire LED substrate. This method is more reproducible and convenient for industrial production, avoiding a series of problems associated with the transfer process. Indeed, the direct deposition of graphene on the surface of nitride semiconductors is the best strategy to integrate the two types of materials in a controllable manner [19]. Meanwhile, using a plasma-enhanced and vertical cold wall CVD system [20–22] not only reduces the growth temperature and protects the material interfaces, it also speeds up the growth, improves the growth efficiency, and reduces the cost as well. Although the quality of a GLTF is yet lower than standard graphene, its properties are good enough to make the GaN LEDs work properly. Through the measurement of electrical, optical, and thermal properties, we found that the addition of GLTFs not only improved the LED luminous current and reduced the turn-on voltage and luminous power consumption, but also had obvious advantages in heat spreading, which has great prospects for improving the reliability, durability and service life of the device. This is one of the key aspects to achieving bright and durable LEDs.

## 2. Experiment

The plasma-enhanced chemical vapor deposition (PECVD) system used in this experiment was a Black Magic Pro system from AIXTRON Nanoinstruments Ltd. (Swavesey, UK). Unlike the traditional tubular furnace growth, a vertical cold wall CVD system was used here, wherein the only heated area was the middle part of the heater while the rest of the area was "cold". Faster growth rate and less energy consumption are the main advantages of this system, a schematic of which is shown in Figure 1a. Figure 1b is the actual growth setup. The graphite heater, which was Joule current heated, can support a maximum 2 inch wafer. Another graphite piece was placed just below the heater, and the direct current (DC) plasma was ignited between these two electrodes. However, our growth parameters were not tuned in favor of vertical graphene formation [23], but rather adjusted towards thin film growth. Figure 1c is a photograph taken during the growth phase (side view, the samples on the heater are seen to be immersed in the glowing plasma). The starting samples used in this study were standard commercial GaN LED epitaxial wafers with sapphire substrates. Using 100 sccm acetylene and 250 sccm argon, GLTF grew on the p-GaN (the outermost layer of the LED wafer), which had been pre-deposited with 2 nm platinum (99.99%), and 40 W-DC plasma was used to reduce the growth temperature because of the fact that temperatures above 600 °C tend to result in dense platinum islands on the surface [24,25]. We grew at 600 °C and 6 mbar (chamber pressure) for 25 min to obtain GLTF directly on p-GaN. The graphene deposition mechanism was primarily plasma-enhanced pyrolysis of hydrocarbons, together with some degree of Pt catalysis during the hydrocarbon decomposition and subsequent graphitization. We prepared GaN LEDs on four types of substrates (denoted by samples 1–4), which were: (1) GaN/sapphire LED substrates with no thermal treatment; (2) 2 nm Pt coated GaN/sapphire with no other treatment; (3) 2 nm Pt coated GaN/sapphire annealed at 600 °C for 25 min; and (4) 25 min of GLTF growth at 600 °C on 2 nm Pt coated GaN/sapphire. The individual LED device had a 260 × 515 μm$^2$ mesa pattern and was fabricated by two steps of photolithography. The manufacturing schematic of the LED devices is shown in Figure 2. First, 120 nm nickel (99.99%)

was deposited on the surface of GLTF as a dry etching mask. After ultraviolet (UV) exposure, high concentration iron trichloride solution was used to remove the unneeded nickel. Afterwards, the wafers with the patterned Ni mask atop were put into an inductively coupled plasma (ICP) dry etching system. The gallium nitride epilayers were etched to a depth of 1.2 μm with a ratio of 8:64 in the $SiCl_4$ and $Cl_2$ gas mixture to reach the heavily doped n-GaN layer, forming the mesa arrays. The surface GLTF and ultrathin platinum were etched and patterned together. After mesa fabrication, iron trichloride was used to remove the rest of the nickel. Finally, Ti/Au (15 nm/300 nm) p and n metal electrodes were fabricated together by lift-off lithography and sputtering. No annealing was conducted in the metal contacts. For a comparison study, GaN LED devices without GLTF were also fabricated with a similar process (samples 2 and 3).

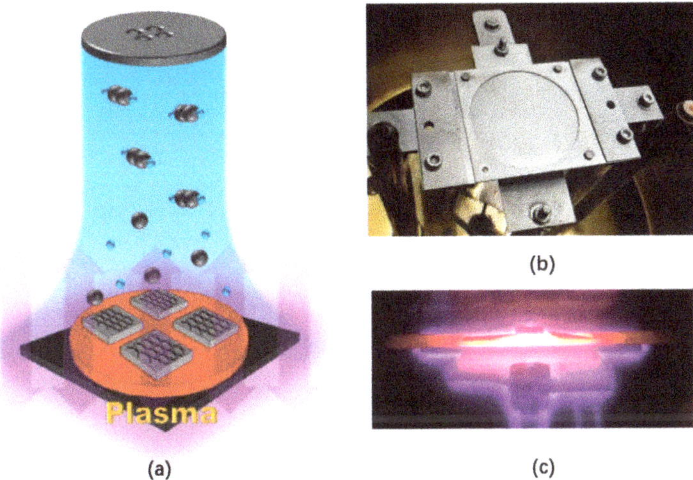


**Figure 1.** (**a**) The graphene growth schematic. The gas ratio was acetylene:argon = 100:250. We used 40 W-direct current (DC) plasma to assist the growth, which occurred over 25 min at 600 °C and 6 mbar. (**b**) The actual setup of the growth chamber, where the plasma was ignited between the graphitic heater and another graphitic electrode beneath. The round area in the middle of the heater is 2 inches in diameter. (**c**) A photo taken during plasma-enhanced chemical vapor deposition (PECVD), showing the glow map.

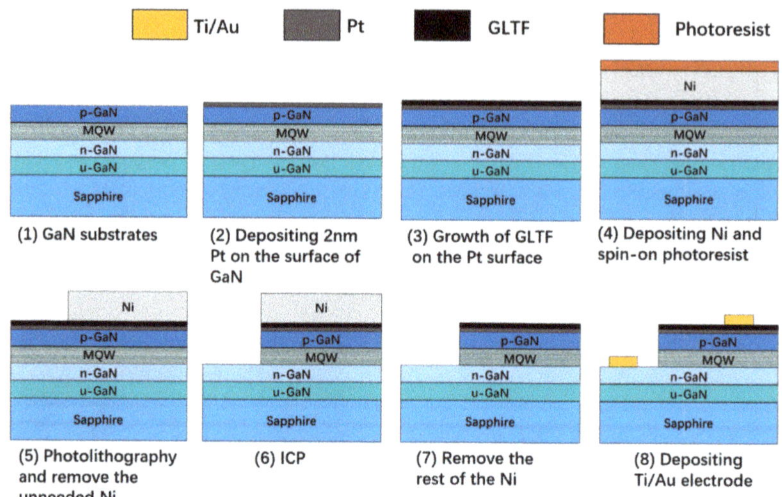


**Figure 2.** The manufacturing schematic of the light-emitting diodes (LEDs). A device was accomplished by depositing 120 nm of Ni as a dry etching mask on a gallium nitride (GaN) sample pre-deposited with 2 nm Pt and graphene-like thin film (GLTF). Photolithography was then conducted and the Ni patterned. After etching into the heavily doped n-GaN layer using an inductively coupled plasma (ICP) system, the rest of the Ni was removed. Ti/Au (15 nm/300 nm) p and n metal electrodes were also fabricated together by lift-off lithography and sputtering. MQW represents a multiple quantum well.

## 3. Results and Discussion

In order to determine whether or not the platinum was affected by the high temperature process, gallium nitride samples with a 2 nm platinum deposit were annealed for 25 min in an argon environment at 600 °C and 40 W-DC plasma. These conditions were nominally the same as the conditions for GLTF growth, except that no carbon source was added. Photographs of the four samples prepared for this study are shown in Figure 3a, labeled as samples 1–4: (1) gallium nitride substrates without any treatment; (2) 2 nm Pt coated p-GaN with no other treatment; (3) 2 nm Pt coated p-GaN in a 600 °C and 40 W DC Ar plasma environment annealed for 25 min; and (4) 25 min of GLTF growth at 600 °C on 2 nm Pt coated p-GaN. For samples 2, 3 and 4, the corresponding sheet resistances were 500, 536 and 472 Ω/sq, respectively. After annealing, the sheet resistance of platinum becomes larger. We noted that the high temperature caused an aggregation effect and, to some extent, turned the platinum film into an island-like structure. Also, the platinum atoms may be partially infiltrated into the p-GaN, further lowering the conductivity of the outer surface. It was clear that the surface roughness of the platinum increased after high temperature annealing. However, after adding the carbon source, the GLTF improved the conductivity of the sample surface despite there still being a high temperature process. Figure 3b shows morphological scanning electron microscope (SEM) images of the four samples. The measured transmittance of GaN LEDs with and without GLTF is plotted as a function of wavelength in Figure 3c. As depicted in the figure, the GaN samples with 2 nm platinum (sample 2) and platinum-GLTF (sample 4) had transparencies in the visible light band of 70–95% and 45–80%, respectively. Both the 2 nm Pt and GLTF reduced the optical transmittance quite a lot. Figure 3d compares the Raman signal of sample 2 and sample 4 at 532 nm wavelength laser excitation. It can be seen that after growing GLTF, two new peaks appear near 1356 $cm^{-1}$ and 1601 $cm^{-1}$, corresponding to the D and G peaks, respectively, which are the signatures of $sp^2$-C graphitic carbon. The D peak is relatively big, indicating that the material grown in this manner contains a number of defects or disorder, which is a subject of future improvement. The Raman spectrum is not that of standard graphene, but rather a nanographitic structure, and therefore in this paper we term it as GLTF for scientific rigor.

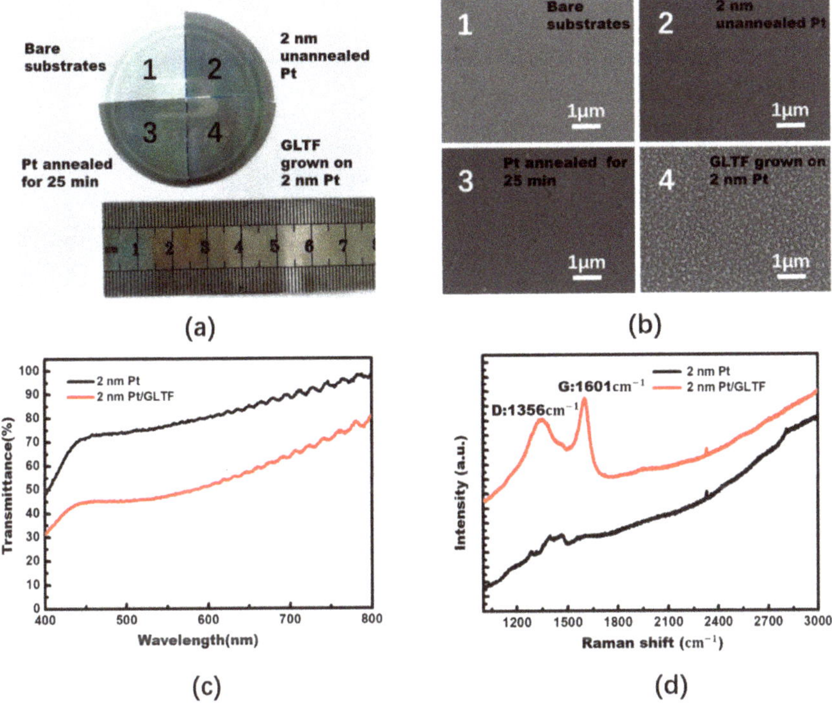


**Figure 3.** (**a**) Shows GaN LED samples of bare substrates, with 2 nm unannealed platinum, with platinum annealed for 25 min, and with GLTF grown on 2 nm platinum (samples 1–4). (**b**) Shows morphological scanning electron microscope (SEM) images of the four samples. It is clear that the platinum surface roughness increased after annealing. (**c**) Shows the transparencies of the GaN LED epiwafers with 2 nm Pt/graphene (sample 4) and 2 nm Pt (sample 2) in the 400–800 nm range. (**d**) Shows a Raman comparison diagram of samples 2 and 4.

As shown in Figure 1, the current–voltage (I–V) characteristics of the samples—GaN substrates without any treatment (sample 1, green line), with 2 nm Pt film (sample 2, black line), with 2 nm annealed Pt film (sample 3, blue line) and with Pt-GLTF film (sample 4, red line)—were compared after making the LED devices. From this comparison, we suggest that, after annealing at 600 °C with 40 W plasma, the performance of the device with 2 nm Pt was worse than its unannealed counterpart. Even at 8 V, the working current was less than 20 mA. The Pt-GLTF device, on the other hand, had a turn-on voltage as low as 3.8 V, while the turn-on voltages of samples 1–3 were about 5.8 V, 4.2 V and 4.5 V, respectively. At 20 mA, sample four had a forward voltage of 3.9 V, and at 5.6 V the current of the device reached 100 mA, which was the limit of our measurement machine. Hence, after adding the carbon source $C_2H_2$ to form GLTF, the performance of the device appeared to be drastically improved under the same operating conditions. In other words, although the Pt film in this device also underwent a high temperature deteriorating process, the addition of GLTF compensated for this effect and made the device outperform the other two types of devices. In fact, we found the turn-on voltage was strongly affected by the sheet resistance and work function of the transparent electrode materials. In sample 1, which did not have a transparent electrode, it was difficult to inject current uniformly and thus it required a large voltage to turn it on. For sample 2, which had the ultrathin Pt film added, its low sheet resistance and relatively good match to p-GaN's Fermi level resulted in a reduction of the turn-on voltage. However, after annealing in sample 3, the Pt became islandic and the sheet resistance (hence the turn-on voltage) went up. The turn-on voltage was brought back down by the addition of GLTF in sample 4, as a result of the reduced sheet resistance.

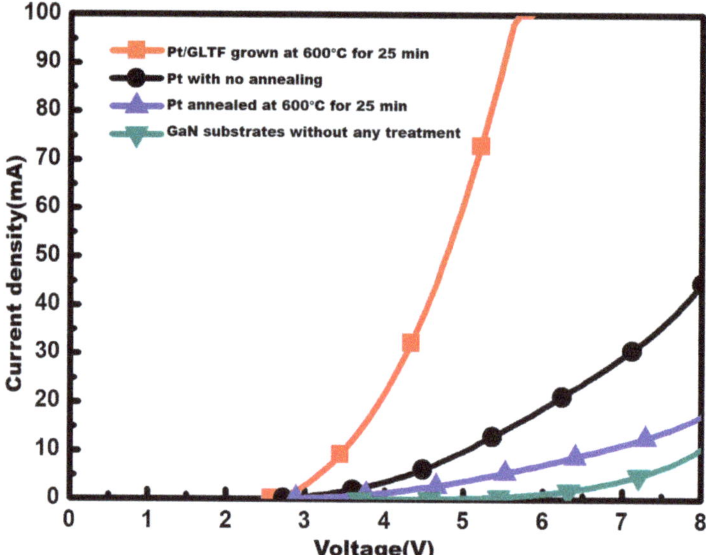


**Figure 4.** A current–voltage (I–V) curve comparison of the LED devices of GaN substrates without any treatment, with Pt film, with annealed Pt film, and with Pt-GLTF film.

Figure 5 displays electroluminescent photos of samples 2–4 operating at 5 V, together with their corresponding optical micrographs for a single device. Figure 5a,b shows the platinum-GLTF LEDs (sample 4), Figure 5c,d shows the platinum film LEDs (sample 2), and Figure 5e,f shows the annealed platinum LEDs (sample 3). It can be seen that the Blu-ray luminescence is very uniform for the platinum-GLTF LED and platinum-LED. The current of the annealed platinum LED at the same voltage is not only smaller, but also unevenly distributed, and the surface of the film looks like it has been damaged. It is proved here that the presence of our directly grown GLTF not only has a very good current spreading effect, but also compensates the negative effects from the Pt annealing, making the current and luminous characteristics of the final LEDs better than other devices.

Figure 6 shows the electroluminescence (EL) spectra of samples 1–4 measured at 20 mA current. The blue emissions of samples 2 and 4 are at 454 and 454.8 nm, respectively, and those of samples 1 and 3 show a redshift of around 10 nm. The exact origin is not clear, possibly related to the quantum-confined Stark effect. The luminous and radiation flux values reflect the electrical to optical energy conversion efficiency (wall-plug efficiency). They were low for sample 1 because without any transparent electrodes, the current injection was very poor and only a spot on the mesa emitted light. The situation is much better for samples 2 and 4, but not for sample 3 since the Pt electrode was damaged by the annealing. For a similar reason, the full width at half maximum for samples 2 and 4 is narrower than sample 3. Sample 1 has the narrowest width—not due to good device performance, but because only one spot was emitting light.

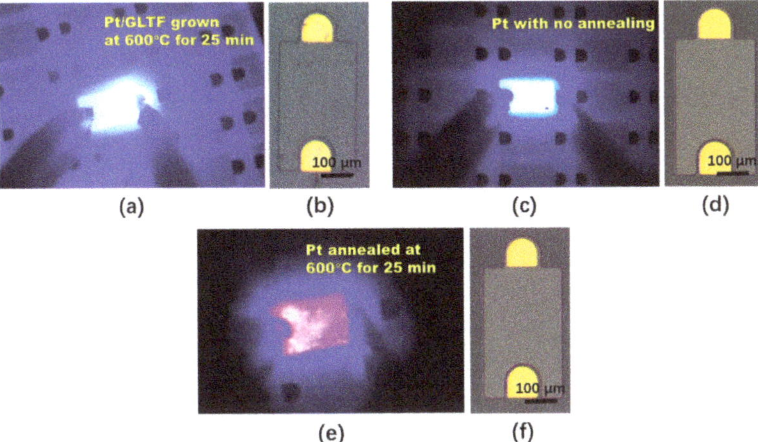


**Figure 5.** (**a**,**b**) Shows the platinum-GLTF LEDs; (**c**,**d**) shows the 2-nm-platinum-film LEDs; and (**e**,**f**) shows the annealed platinum LEDs.

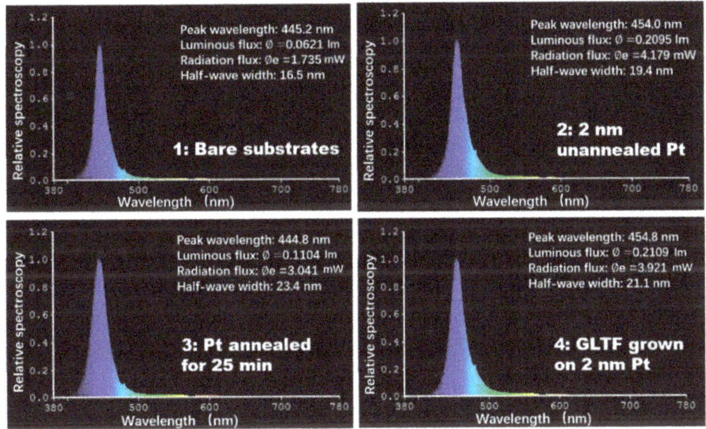


**Figure 6.** Electroluminescence spectra of GaN LED samples 1–4 measured at 20 mA injection current.

Finally, in order to study the thermal management characteristics of the as-grown GLTF, we welded and encapsulated three types of LEDs. The package welding diagram of the GaN LEDs is shown in Figure 7. The pentagon in the middle is the GaN piece, the yellow part at the top is the wire, and the red arrows indicate the zoomed-in parts shown in detail in the insets. The p- and n-poles of each LED were connected to the wires via wire bonding for thermal distribution measurements. Using an SC7300M F/2 (MCT) thermal imaging camera (FLIR SYSTEMS, Wilsonville, OR, USA), accurate temperatures of the three LEDs were measured both at room temperature and in operation (in 20 mA constant current mode or 8 V constant voltage mode).

**Figure 7.** Wire bonding and packaging diagram of the LEDs for thermal measurement. The pentagon in the middle is the GaN piece which was diced to fit the measurement setup.

Figures 8–10 show the platinum-GLTF LEDs (sample 4), platinum film LEDs (sample 2), and platinum-annealed LEDs (sample 3), respectively, under three different conditions for measurement of thermal distribution. In the figures, "px" represents different positions on the film from n to p electrodes along the device (indicated by the solid lines in panels (a–c) in Figures 8–10). Panels (a) and (d) were recorded at room temperature when the device was not operating, (b) and (e) were recorded with a constant working current at 20 mA, and (c) and (f) were recorded with a constant voltage at 8 V. Considering Figure 8, by collecting the highest and the lowest temperatures in the temperature distribution along the crossline, it can be estimated that, compared to room temperature (i.e., no operation) at 20 mA, the temperature increased by 0.7–0.88 °C in platinum-GLTF LEDs. At 8 V voltage, platinum-GLTF LEDs reached the limit current of 100 mA in the measurement system, and were about 12.52 to 17.69 °C higher than room temperature.

As shown in Figure 9, the platinum film LED increased its temperature by 1.5–1.62 °C at 20 mA work current, which was much higher than that of the platinum-GLTF LED, indicating that at the same current injection, the Pt film LED generates a lot more heat. Its temperature increased by 4.62–6.18 °C at 8 V work voltage, lower than the Pt-GLTF LED in Figure 8. This, however, is not because it has superior thermal dissipation. Rather, it is because of its low current, caused by the low current injection performance.

As seen in Figure 10, the temperature of platinum-annealed LEDs increased by 1.59–1.87 °C at 20 mA, while their temperature at 8 V increased by only 1.53–1.78 °C, which may be attributable to the very low current. Based on the data in Figures 8–10, we conclude that, at the same current, platinum-GLTF LEDs require less voltage and hence less input power, generating significantly less heat than both nonannealed and annealed platinum LEDs. At the same working voltage, the input power of the Pt-GLTF LEDs is very large because of its large current, which is in turn the result of the good injection performance. Therefore, the temperatures associated with the Pt-GLTF devices are naturally higher than the other two types of devices. Importantly, a heat spreading effect was observed in sample 4. For example, Figure 8c shows the hot spot on the chip is around the p electrode and there is clear spreading of the heat towards the n electrode side across the mesa. A similar effect can be seen in Figure 8b. In contrast, for sample 2 in Figure 9c, the heat is strictly concentrated around the p electrode. Even the trenches between mesas near the p side became hot, but still the heat could not spread towards the n electrode along the mesa. Heat spreading was out of the question for sample 3, as poor injection meant its current was too small to generate any hot spots. Our results show that the heat spreading of GLTF is good, thanks to its very high thermal conductivity.

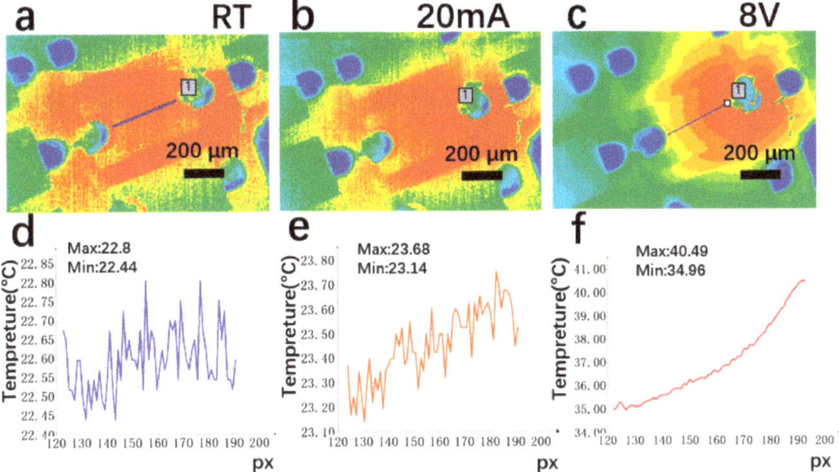


**Figure 8.** Temperature distribution of platinum-GLTF LEDs (sample 4). In (**b**,**c**), the heat is seen to spread from the p electrode side towards the n electrode side via the GLTF-based transparent electrode. Here, (**a**,**d**) represent room temperature with no operation; (**b**,**e**) represent operation at a constant current of 20 mA; and (**c**,**f**) represent constant voltage working mode at 8 V.

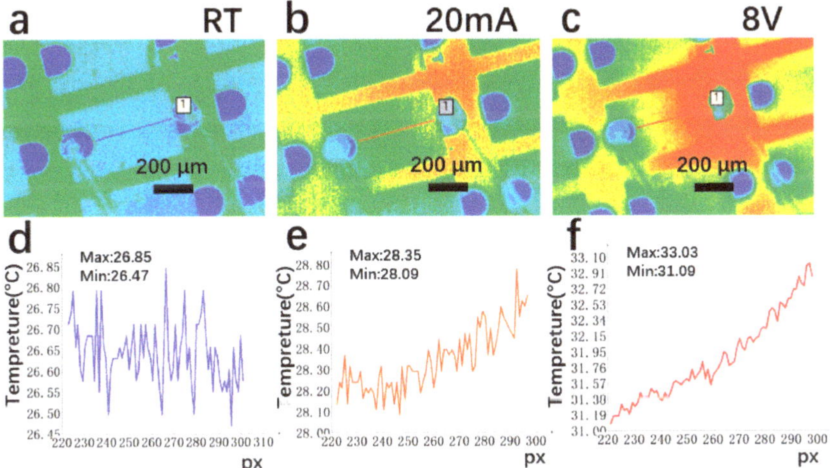


**Figure 9.** Temperature distribution of platinum film LEDs (sample 2). In (**c**), the heat is accumulated around the p electrode and the nearby trenches and cannot be effectively dissipated or spread. (**a**,**d**): room temperature, no operation; (**b**,**e**): working at constant current 20 mA; (c,**f**): constant voltage (8 V) operation.

The overall performance of the devices prepared by our method is reasonably good. However, at this stage, it is still not able to compete with commercial GaN LEDs. For example, the work voltage of standard blue LEDs at 20 mA is only slightly above 3 V. Despite this, compared to graphene-based devices, our devices are very competitive with a 3.9 V work voltage at 20 mA. Commonly, graphene-on-GaN LEDs have forward work voltages of 6–7 V or even more than 10 V [26,27].

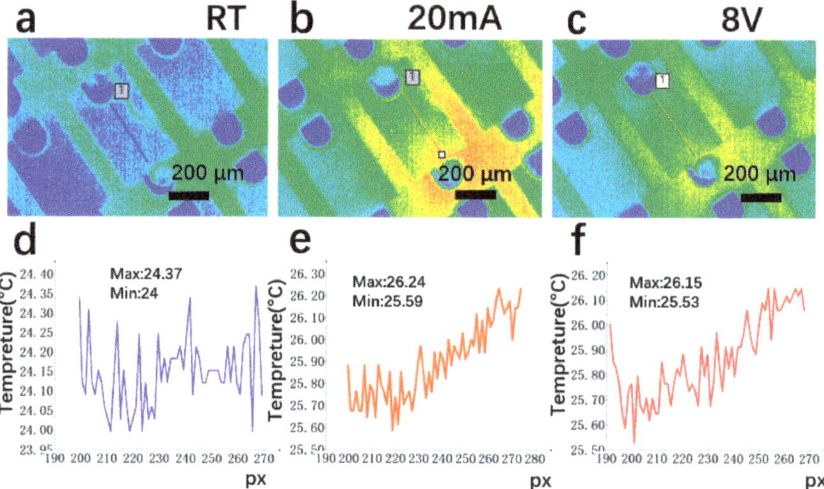


**Figure 10.** Temperature distribution of platinum-annealed LEDs (sample 3). (**a**,**d**): room temperature; (**b**,**e**): constant current at 20 mA; and (**c**,**f**): constant voltage at 8 V.

## 4. Conclusions

Based on graphene's high electrical and thermal conductivity, high optical transmittance, rich raw materials, broad spectrum and good chemical stability, there have been a considerable number of papers suggesting graphene is a promising candidate for substitution of ITO transparent electrodes in GaN-based optoelectronics. However, as early as 2012 [28], we recognized the importance of removing the graphene transfer process, because otherwise the technology would conceivably receive no real interest from the semiconductor industry. That is because graphene transfer is tricky, time consuming and irreproducible. In other words, it is incompatible with semiconductor processing, leaving holes, wrinkles and etching residues in the graphene and its interface, as well as damaging the material quality and the graphene-GaN contact. Thus, in this paper, a method of direct growth of GLTF on ultrathin platinum on GaN was demonstrated to avoid transfer. Unlike inert substrates such as $SiO_2$, GaN is relatively prone to damage and requires subtle control over the growth process [29,30]. In our case, the growth temperature was reduced by the plasma enhancement technique, and the vertical cold wall system reduced the deposition time. In this way, the exposure of the GaN wafer to high temperatures was limited and the GaN was intact (free from surface decomposition). The 2 nm Pt used in this study helped to catalyze the growth and yield of GLTF, producing results that were better than previous attempts which did not use a catalyst [28], yet was thin enough to let light pass through. Through a comparison study of samples with and without GLTF, we found that, compared to platinum-only LEDs (annealed and nonannealed), the surface sheet resistance and the turn-on voltage of the GLTF devices were smaller. At the same current, the platinum-GLTF LEDs required less voltage and hence less luminous power consumption. In the heat spreading characterization, the GLTF devices were also superior, thanks to the intrinsically high thermal conductivity of GLTF. At 20 mA, the temperature of platinum-GLTF LEDs increased by less than 1 °C, significantly lower than the control samples. Therefore, the as-prepared GLTF-based transparent electrodes had better current spreading, current injection and heat spreading functions in the GaN LEDs. We also identified high temperature annealing as a deteriorating process of 2 nm Pt. However, adding GLTF on top of Pt compensated for this effect and led to better performance. This work confirms that as-grown, transfer-free GLTF has clear advantages in developing high performance, scalable, controllable and reproducible transparent electrodes for GaN-based optoelectronics. At this stage, however, optical transmittance remains a challenge. Seeking an ultrathin metal catalyst with better transparency than Pt and optimization of full-scale growth parameters are expected to refine the technology to a level that is suitable for real industrial applications.

**Author Contributions:** Data curation, F.X.; Formal analysis, X.L., Z.D. and L.W.; Methodology, W.G.; Resources, S.F. and J.D.; Supervision and methodology, J.S.

**Funding:** This work was supported by the National Key Research and Development Program of China under Grant 2017YFB0403100 and Grant 2017YFB0403102, and the National Natural Science Foundation of China under Grant 11674016.

**Conflicts of Interest:** The authors declare no conflict of interest.

## References

1. Kung, P.; Saxler, A.; Zhang, X.; Walker, D.; Wang, T.C.; Ferguson, I.; Razeghib, M. High quality AlN and GaN epilayers grown on (00·1) sapphire, (100), and (111) silicon substrates. *Appl. Phys. Lett.* **1995**, *66*, 2958–2960. [CrossRef]
2. Morkoç, H.; Strite, S.; Gao, G.B.; Lin, M.E.; Sverdlov, B.; Burns, M. Large-band-gap SiC, III-V nitride, and II-VI ZnSe-based semiconductor device technologies. *J. Appl. Phys.* **1994**, *76*, 1363–1398. [CrossRef]
3. Fasol, G. Room-temperature blue gallium nitride laser diode. *Science* **1996**, *272*, 1751–1752. [CrossRef]
4. Nakamura, S. The roles of structural imperfections in InGaN-based blue light-emitting diodes and laser diodes. *Science* **1998**, *281*, 956–961. [CrossRef]
5. Hamberg, I.; Granqvist, C.G. Evaporated Sn-doped $In_2O_3$ films: Basic optical properties and applications to energy-efficient windows. *J. Appl. Phys.* **1986**, *60*, 123–160. [CrossRef]
6. Ellmer, K. Past achievements and future challenges in the development of optically transparent electrodes. *Nature Photon.* **2012**, *6*, 809–817. [CrossRef]
7. Xu, K.; Xu, C.; Xie, Y.; Deng, J.; Zhu, Y.; Guo, W.; Mao, M.; Xun, M.; Chen, M.; Zheng, L.; et al. GaN nanorod light emitting diodes with suspended graphene transparent electrodes grown by rapid chemical vapor deposition. *Appl. Phys. Lett.* **2013**, *103*, 222105. [CrossRef]
8. Novoselov, K.S. Electric field effect in atomically thin carbon films. *Science* **2004**, *306*, 666–669. [CrossRef]
9. Ding, X.; Sun, H.; Gu, X. The direct synthesis of graphene on a gallium nitride substrate. *Chem. Vap. Depos.* **2014**, *20*, 125–129. [CrossRef]
10. Jo, G.; Choe, M.; Cho, C.Y.; Kim, J.H.; Park, W.; Lee, S.; Hong, W.K.; Kim, T.W.; Park, S.J.; Hong, B.H.; et al. Large-scale patterned multi-layer graphene films as transparent conducting electrodes for GaN light-emitting diodes. *Nanotechnology* **2010**, *21*, 175201. [CrossRef]
11. Kim, B.J.; Mastro, M.A.; Hite, J.K.; Eddy, C.R.; Kim, J. Transparent conductive graphene electrode in GaN-based ultra-violet light emitting diodes. *Opt. Express* **2010**, *18*, 23030. [CrossRef]
12. Kim, B.J.; Lee, C.; Jung, Y.; Baik, K.H.; Mastro, M.A.; Hite, J.K.; Eddy, C.R.; Kim, J. Large-area transparent conductive few-layer graphene electrode in GaN-based ultra-violet light-emitting diodes. *Appl. Phys. Lett.* **2011**, *99*, 143101. [CrossRef]
13. Gao, L.; Ren, W.; Xu, H.; Jin, L.; Wang, Z.; Ma, T.; Ma, L.P.; Zhang, Z.; Fu, Q.; Peng, L.M.; et al. Repeated growth and bubbling transfer of graphene with millimetre-size single-crystal grains using platinum. *Nat. Commun.* **2012**, *3*, 699. [CrossRef]
14. Ping, J.; Fuhrer, M.S. Layer Number and Stacking Sequence Imaging of Few-Layer Graphene by Transmission Electron Microscopy. *Nano Lett.* **2012**, *12*, 4635–4641. [CrossRef] [PubMed]
15. Sutter, P.; Sadowski, J.T.; Sutter, E. Graphene on Pt (111): Growth and substrate interaction. *Phys. Rev. B* **2009**, *80*, 245411. [CrossRef]
16. Nam, Y.; Sun, J.; Lindvall, N.; Yang, J.S.; Kireev, D.; Park, R.C.; Park, Y.W.; Yurgens, A. Quantum hall effect in graphene decorated with disordered multilayer patches. *Appl. Phys. Lett.* **2013**, *103*, 233110. [CrossRef]
17. Shi, R.; Xu, H.; Chen, B.; Zhang, Z.; Peng, L.M. Scalable fabrication of graphene devices through photolithography. *Appl. Phys. Lett.* **2013**, *102*, 113102. [CrossRef]
18. Sun, J.; Nam, Y.; Lindvall, N.; Cole, M.T.; Teo, K.B.K.; Park, Y.W.; Yurgens, A. Growth mechanism of graphene on platinum: Surface catalysis and carbon segregation. *Appl. Phys. Lett.* **2014**, *104*, 152107. [CrossRef]

19. Giannazzo, F.; Fisichella, G.; Greco, G.; La Magna, A.; Roccaforte, F.; Pecz, B.; Yakimova, R.; Dagher, R.; Michon, A.; Cordier, Y. Graphene integration with nitride semiconductors for high power and high frequency electronics. *Phys. Status Solidi A* **2017**, *214*, 1600460. [CrossRef]
20. Sun, J.; Lindvall, N.; Cole, M.T.; Angel, K.T.T.; Wang, T.; Teo, K.B.K.; Chua, D.H.C.; Liu, J.; Yurgens, A. Low partial pressure chemical vapor deposition of graphene on copper. *IEEE Trans. Nanotechnol.* **2012**, *11*, 255–260. [CrossRef]
21. Sun, J.; Lindvall, N.; Cole, M.T.; Teo, K.B.K.; Yurgens, A. Large-area uniform graphene-like thin films grown by chemical vapor deposition directly on silicon nitride. *Appl. Phys. Lett.* **2011**, *98*, 252107.
22. Sun, J.; Cole, M.T.; Lindvall, N.; Teo, K.B.K.; Yurgens, A. Noncatalytic chemical vapor deposition of graphene on high-temperature substrates for transparent electrodes. *Appl. Phys. Lett.* **2012**, *100*, 022102.
23. Ghosh, S.; Polaki, S.R.R.; Kamruddin, M.; Jeong, S.M.; Ostrikov, K. Plasma-electric field controlled growth of oriented graphene for energy storage applications. *J. Phys. D Appl. Phys.* **2018**, *51*, 145303. [CrossRef]
24. Duxstad, K.J.; Haller, E.E.; Yu, K.M. High temperature behavior of Pt and Pd on GaN. *J. Appl. Phys.* **1997**, *81*, 3134. [CrossRef]
25. Kim, H.G.; Deb, P.; Sands, T. High-reflectivity Al-Pt nanostructured Ohmic contact to p-GaN. *IEEE Trans. Electron. Devices* **2006**, *53*, 2448–2453.
26. Shim, J.-P.; Seo, T.H.; Min, J.-H.; Kang, C.M.; Suh, E.-K.; Lee, D.-S. Thin Ni film on graphene current spreading layer for GaN-based blue and ultra-violet light-emitting diodes. *Appl. Phys. Lett.* **2013**, *102*, 151115. [CrossRef]
27. Kun, X.; Chen, X.; Jun, D.; Yanxu, Z.; Weiling, G.; Mingming, M.; Lei, Z.; Jie, S. Graphene transparent electrodes grown by rapid chemical vapor deposition with ultrathin indium tin oxide contact layers for GaN light emitting diodes. *Appl. Phys. Lett.* **2013**, *102*, 162102. [CrossRef]
28. Sun, J.; Cole, M.T.; Ahmad, S.A.; Backe, O.; Ive, T.; Loffler, M.; Lindvall, N.; Olsson, E.; Teo, K.B.K.; Liu, J.; et al. Direct chemical vapor deposition of large-area carbon thin films on gallium nitride for transparent electrodes: A first attempt. *IEEE Trans. Semicond. Manuf.* **2012**, *25*, 494–501. [CrossRef]
29. Parida, S.; Magudapathy, P.; Sivadasan, A.K.; Pandian, R.; Dhara, S. Optical properties of AlGaN nanowires synthesized via ion beam techniques. *J. Appl. Phys.* **2017**, *121*, 205901. [CrossRef]
30. Parida, S.; Patsha, A.; Bera, S.; Dhara, S. Spectroscopic investigation of native defect induced electron–phonon coupling in GaN nanowires. *J. Phys. D Appl. Phys.* **2017**, *50*, 275103. [CrossRef]

*Article*

# Direct Growth of AlGaN Nanorod LEDs on Graphene-Covered Si

**Fang Ren [1,2,3], Yue Yin [1,2,3], Yunyu Wang [1,2,3], Zhiqiang Liu [1,2,3,*], Meng Liang [1,2,3], Haiyan Ou [4], Jinping Ao [5], Tongbo Wei [1,2,3], Jianchang Yan [1,2,3], Guodong Yuan [1,2,3,*], Xiaoyan Yi [1,2,3,*], Junxi Wang [1,2,3], Jinmin Li [1,2,3], Dheeraj Dasa [6] and Helge Weman [6,7,*]**

1 Research and Development Center for Solid State Lighting, Institute of Semiconductors, Chinese Academy of Sciences, Beijing 100083, China; rf@semi.ac.cn (F.R.); yinyue@semi.ac.cn (Y.Y.); wyyu@semi.ac.cn (Y.W.); liangmeng@semi.ac.cn (M.L.); tbwei@semi.ac.cn (T.W.); yanjc@semi.ac.cn (J.Y.); jxwang@red.semi.ac.cn (J.W.); jmli@red.semi.ac.cn (J.L.)
2 Center of Materials Science and Optoelectronics Engineering, University of Chinese Academy of Sciences, Beijing 100049, China
3 Beijing Engineering Research Center for the 3rd Generation Semiconductor Materials and Application, Beijing 100083, China
4 Department of Photonics Engineering, Technical University of Denmark, Ørsteds Plads 345A, DK-2800 Kongens Lyngby, Denmark; haou@fotonik.dtu.dk
5 Department of Electrical and Electronic Engineering, The University of Tokushima, 2-1, Minami-josanjima, Tokushima 770-8506, Japan; jpao@xidian.edu.cn
6 CrayoNano AS, Sluppenvegen 6, NO-7037 Trondheim, Norway; dheeraj.dasa@crayonano.com
7 Department of Electronics and Telecommunications, Norwegian University of Science and Technology (NTNU), NO-7491 Trondheim, Norway
* Correspondence: lzq@semi.ac.cn (Z.L.); gdyuan@semi.ac.cn (G.Y.); spring@semi.ac.cn (X.Y.); helge.weman@ntnu.no (H.W.); Tel.: +86-010-8230-5423 (Z.L.)

Received: 9 November 2018; Accepted: 23 November 2018; Published: 26 November 2018

**Abstract:** High density of defects and stress owing to the lattice and thermal mismatch between nitride materials and heterogeneous substrates have always been important problems and limit the development of nitride materials. In this paper, AlGaN light-emitting diodes (LEDs) were grown directly on a single-layer graphene-covered Si (111) substrate by metal organic chemical vapor deposition (MOCVD) without a metal catalyst. The nanorods was nucleated by AlGaN nucleation islands with a 35% Al composition, and included n-AlGaN, 6 period of AlGaN multiple quantum wells (MQWs), and p-AlGaN. Scanning electron microscopy (SEM) and electron backscatter diffraction (EBSD) showed that the nanorods were vertically aligned and had an accordant orientation along the [0001] direction. The structure of AlGaN nanorod LEDs was investigated by scanning transmission electron microscopy (STEM). Raman measurements of graphene before and after MOCVD growth revealed the graphene could withstand the high temperature and ammonia atmosphere in MOCVD. Photoluminescence (PL) and cathodoluminescence (CL) characterized an emission at ~325 nm and demonstrated the low defects density in AlGaN nanorod LEDs.

**Keywords:** AlGaN; nanorod LEDs; graphene; MOCVD

---

## 1. Introduction

GaN-based semiconductor materials including AlGaN and InGaN have been considered as ideal materials for light emitting diodes (LED), laser diodes (LD), solar cells, and other optoelectronic devices on account of their direct bandgap characteristics and adjustable bandgap width. However, due to the lack of large-scale readily available single crystal substrate, GaN-based materials are usually heteroepitaxial by metal organic chemical vapor deposition (MOCVD) or molecular beam epitaxy (MBE) on sapphire, Si, or SiC, which have large lattice and thermal expansion mismatch.

To avoid defects such as dislocations or stacking faults caused by strain during the epitaxial process, a novel approach is proposed involving the growth of nanostructures such as nanorods, which have a high aspect-ratio and large surface-to-volume ratio, consequently releasing the strain and reducing the dislocation density in the upper part of the nanorods [1,2]. Moreover, core/shell nanorod structures are reported to be beneficial to increase the overall area of emission via the regrowth of the active region shell on the nanorod core, resulting in the improvement of total light intensity of the same substrate area [3], as well as avoiding the strong spontaneous and piezoelectric polarization fields through the high quality m-plane nonpolar facets [4]. Coulon et al. [5,6] reported the fabrication of core-shell LED structures using an original hybrid top-down/bottom-up approach, which achieved emission at the deep ultraviolet band. Zhuang et al. [7] developed a soft UV-curing nanoimprint lithography (NIL) technique for fabricating GaN nanogratings and nanorods, followed by reactive ion etching (RIE) and inductively coupled plasma (ICP) system processes. Our group also reported the controlled growth of GaN nanowires by a metal-catalyzed method using hydride vapor phase epitaxy (HVPE) system and achieved the orientation-controllable GaN nanowires with a high aspect ratio and excellent crystal quality [8,9].

However, the above methods for nanorods needed an original AlN or GaN template or metal catalyst—which complicated the fabrication of nanostructures. Graphene, a two-dimensional planar configuration of $sp^2$-bonded carbon atoms, has attracted a great interest owing to the hexagonal arrangement of C atoms, making the one-atomic layer graphene able to serve as a nearly lattice-matched buffer for the growth of wurtzite GaN. The GaN-based nanorods can be directly grown on graphene-covered substrates without the need of a crystalline bulk or metal catalyst. Besides, graphene has other excellent physical and chemical properties, such as high optical transparency, low electrical resistivity, and mechanical strength and flexibility [10–12]. Furthermore, graphene films are transferable to almost any carrier substrate, including amorphous and flexible materials [13]. Therefore, the growth of GaN-based nanorods on a graphene buffer provides a new idea for fabrication of flexible optoelectrical devices. Chung et al. [14] reported the fabrication of bendable LED using high-quality GaN microdisks grown on patterned graphene microdots by epitaxial lateral overgrowth (ELOG). Heilmann et al. [15] demonstrated c-axis-oriented growth of vertically aligned GaN nanorods using single-layer graphene as an atomically thin buffer layer. Kumaresan et al. [13] reported epitaxial growth of defect-free GaN nanowires on graphene using molecular beam epitaxy without any catalyst or intermediate layer. In our previous work, we studied the direct growth of high-quality AlN films on graphene buffer, and the XRD showed the full width at half maximum (FWHM) values for (0002) and $(10\bar{1}2)$ reflections were 360 and 622.2 arcsec, respectively, which were lower than that of the film directly grown on sapphire [16,17]. Moreover, we also demonstrated the GaN-based LEDs grown on multilayer graphene, which showed a higher output power than those grown on conventional sapphire [10].

In this work, we demonstrated a self-organized growth of AlGaN nanorods with a full LED structure on single-layer graphene-covered Si substrate without a metal catalyst by MOCVD. The nanorods was nucleated by AlGaN nucleation islands with a high Al composition, and grew vertically with [0001] orientation. The morphology, orientation, crystal structure, and optical properties were analyzed.

## 2. Materials and Methods

The key processes involved in the graphene wet-transfer procedure and the MOCVD growth of AlGaN nanorods are schematically shown in Figure 1. Prior to MOCVD growth of AlGaN nanorod LEDs, the single-layer graphene film grown on Cu foil by atmospheric-pressure chemical vapor deposition (APCVD, Xicheng, Xiamen, China) was transferred onto a Si (111) substrate. To ensure the high quality of graphene film, poly(methyl methacrylate) (PMMA) was spin-coated onto the Cu foil and baked at 120 °C for 15 min. Then, the Cu foil was immersed into an aqueous solution of iron trichloride ($FeCl_3$) for 4 h to dissolve the Cu substrate entirely. After that, the graphene with PMMA needed to be transferred into deionized water two or three times to wash away residual $FeCl_3$. Then, the graphene was transferred onto cleaned Si (111) substrate and dried in nitrogen. Finally, PMMA was removed using acetone and ethanol.

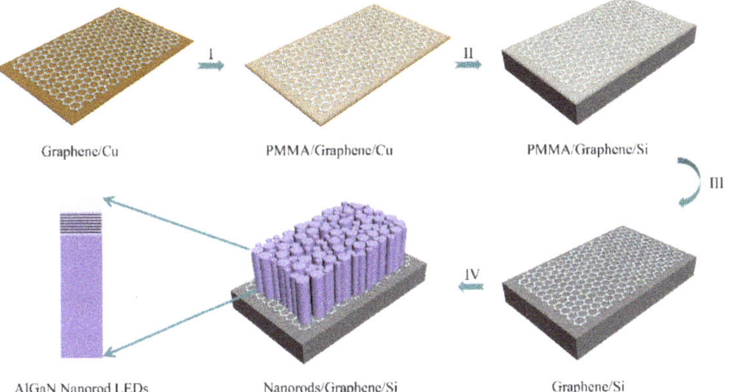


**Figure 1.** Schematic diagram of the key processes involved in the graphene transfer procedure and the metal organic chemical vapor deposition (MOCVD) growth of AlGaN nanorod LEDs: (**I**) Spin-coated poly(methyl methacrylate) (PMMA) onto graphene on Cu foil; (**II**) transfer of graphene with PMMA onto Si substrate; (**III**) dissolving PMMA; and (**IV**) MOCVD growth of AlGaN nanorod light emitting diodes (LEDs).

Throughout the growth process, we adopted trimethylgallium (TMGa), trimethylaluminum (TMAl), and ammonia ($NH_3$) as precursors, silane ($SiH_4$) and magnesocene ($Cp_2Mg$) as dopants, and hydrogen ($H_2$) as the carrier gas. The gas flow rate of precursors and dopants for all the MOCVD processes was summarized in Table 1. Before growth initiation, a nitridation step was utilized by introducing $NH_3$ with a flow of 1000 sccm at 1090 °C for 5 min. Due to the higher adsorption energy and lower migration energy barrier on graphene than Ga atom, the use of Al atoms was beneficial for the adsorption at the growth interface without the participant of defects and facilitated the formation of nucleation points which supported the growth of nanorods [18]. We grew n-AlGaN nucleation islands with a relatively high Al component of 35% for 42 s by introducing TMGa and TMAl with flux of 17.5 and 30 sccm, while the flux of $NH_3$ was 1000 sccm. Subsequently, the n-AlGaN nanorods with an Al component of 11% were grown at the same temperature for 25 min by introducing $SiH_4$ into the reactor. The flow of TMGa and TMAl are 35 and 290 sccm respectively. Lin et al. [19,20] demonstrated that a lower V/III molar ratio could increase the vertical-to-lateral aspect ratio, consequently promoting vertical growth of nanorods. Therefore, the $NH_3$ flow was 15 sccm during the whole process of nanorods growth. The MQWs structure contained 6 pairs of undoped $Al_{0.04}Ga_{0.96}N/Al_{0.11}Ga_{0.89}N$, and both the growth time of walls and barriers were 1 min. Finally, a thin Mg doped p-AlGaN layer was grown for 6 min.

**Table 1.** The gas flow rate of precursors and dopants for all the MOCVD processes.

| Step | $NH_3$ (sccm) | TMGa (sccm) | TMAl (sccm) | $SiH_4$ (sccm) | $Cp_2Mg$ (sccm) |
|---|---|---|---|---|---|
| Nitridation | 1000 | / | / | / | / |
| n-AlGaN | 1000 | 17.5 | 30 | 500 | / |
| nucleation islands | 15 | 35 | 290 | 500 | / |
| n-AlGaN nanorods | 15 | 35 | 250 | / | / |
| u-AlGaN MQWs | 15 | 35 | 290 | / | / |
| p-AlGaN | 15 | 30 | 250 | / | 150 |

The morphology, orientation, and crystal structure of the AlGaN nanorod LEDs were characterized by scanning electron microscopy (SEM, Hitachi, Tokyo, Japan), electron backscatter diffraction (EBSD, Zeiss, Jena, Germany) and scanning transmission electron microscopy (STEM, Tecnai, Hillsboro, OR, USA), respectively. The Raman spectra (Horiba, Kyoto, Japan) of graphene on Si substrate before and after MOCVD growth were collected using a 532-nm laser, which was excited using an argon ion laser. At last, the optical properties including temperature dependent photoluminescence (PL) and cathodoluminescence (CL) mappings of nanorods were analyzed.

## 3. Results and Discussion

The morphology of AlGaN nanorod LEDs grown on the graphene was investigated by scanning electron microscopy (SEM). Figure 2a,b showed the 25° tilted-view SEM image and cross-sectional SEM image of the AlGaN nanorod LEDs grown on graphene. The nanorods were vertically aligned and had a uniform height of $440 \pm 10$ nm and diameter of $200 \pm 10$ nm. In order to investigate the grown orientation of nanorods, we further obtained the electron backscatter diffraction (EBSD) inverse pole figure (IPF). In the wurtzite nitride materials, the (0001) plane usually referred to a plane composed of alternating diatomic close packed of Ga/Al and N pairs, and the [0001] direction, also called c-axis orientation, was perpendicular to the (0001) plane. In term of MOCVD, nitrides materials were usually grown along c-axis orientation. Figure 2c showed the normal-direction IPF image of AlGaN nanorods. It was obvious that almost all AlGaN nanorods exhibited red color in the normal-direction, which demonstrated that the AlGaN nanorods had an accordant orientation along the c-axis direction. Nevertheless, blue and green colors were observed in the transverse-direction IPF image, as shown in Figure 2d. The variations of the inplane orientation of AlGaN nanorods may be due to the random growth of AlGaN nucleation islands on the graphene with different orientation.

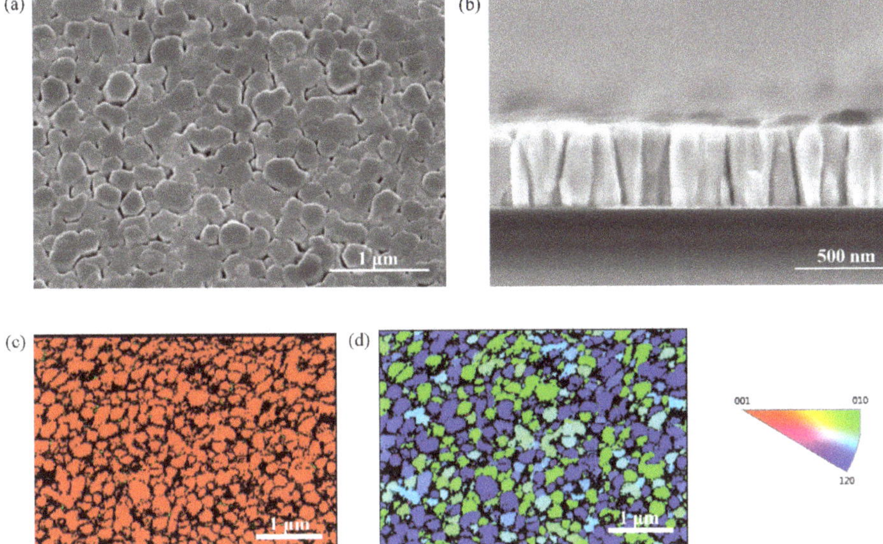


**Figure 2.** The morphology and orientation of AlGaN nanorod LEDs: (**a**) The 25° tilted-view SEM image of AlGaN nanorod LEDs; (**b**) the cross-sectional SEM image of AlGaN nanorod LEDs; (**c**) the normal-direction electron backscatter diffraction (EBSD) inverse pole figure (IPF) image of AlGaN nanorod LEDs; (**d**) the transverse-direction EBSD IPF image of AlGaN nanorod LEDs.

We further investigated the structural characteristics of the GaN nanorod LEDs by STEM using a high angle annular dark field (HAADF) detector. The STEM acceleration voltage was set to 300 kV to analyze the nanostructure. Some bright spots could be observed in Figure 3, especially in Figure 3b, which were metal precipitations introduced during the sample preparation process by Focused Ion Beam (FIB). Figure 3a showed the cross-sectional STEM image of a single AlGaN nanorod LED. From bottom to top, three parts could be observed obviously including n-AlGaN region, MQWs active region, and p-AlGaN layer. Figure 3b showed the amplified MQWs structures. Six pairs of $Al_{0.04}Ga_{0.96}N/Al_{0.11}Ga_{0.89}N$ MQWs with a uniform thickness of 8 nm could be observed with abrupt interfaces. At the top edge of nanorods, as shown in Figure 3c, the MQWs and p-AlGaN was curving downward, which was like a core-shell structure. However, due to a high density of AlGaN nanorods, the core-shell structure was not formed completely.

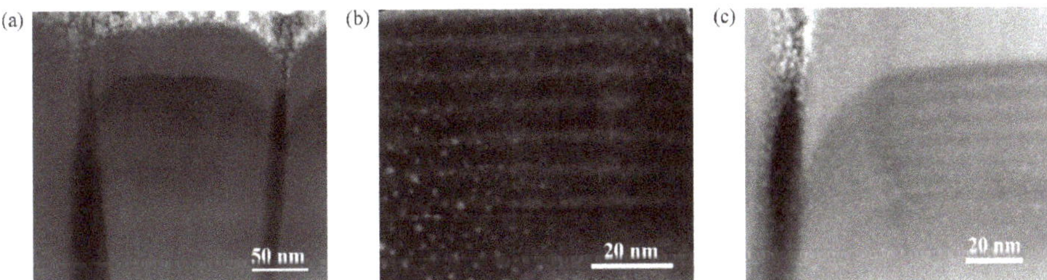


**Figure 3.** The structure of AlGaN nanorod LEDs: (**a**) cross-sectional STEM image of a single AlGaN nanorod LED; (**b**) the amplified MQWs structure; and (**c**) the curved downward MQWs and p-AlGaN.

Figure 4 summarized the Raman spectra of graphene's responses before and after MOCVD growth. The G and 2D peaks (~1580 and ~2700 $cm^{-1}$, respectively) could be observed clearly in both spectra, which demonstrated that graphene could withstand the severe growth conditions for AlGaN nanorod

LEDs in MOCVD. However, compared with the spectrum before MOCVD growth, the D peak, at 1354 $cm^{-1}$, was visible clearly after growth, illustrating the formation of defects during the process of growth on account of the etching by ammonia. These defects could increase the resistivity of grapheme [21]. Furthermore, before the MOCVD growth, the line-shape of 2D peak was systematical and the ratio of the intensity of the 2D peak to the G peak was about 2.3, which revealed the characteristic for single-layer grapheme [22,23]. After MOCVD growth, the G and 2D peaks both shifted to higher frequencies, attributed to nitrogen doping of graphene and the compressive strain during MOCVD growth, especially in the nitridation step [24].

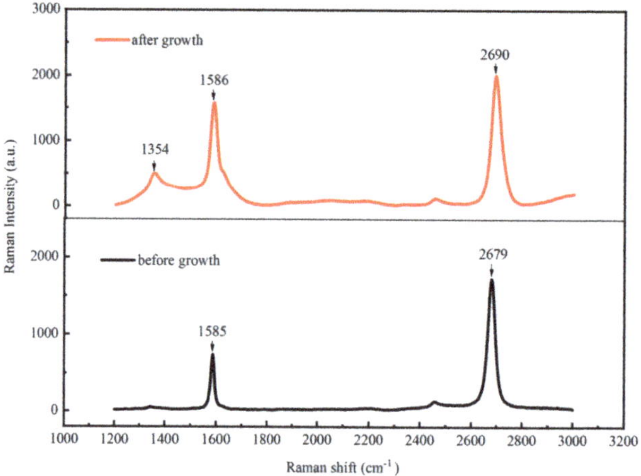


**Figure 4.** The Raman spectra of graphene's responses before and after MOCVD growth.

We measured the photoluminescence (PL) spectra of AlGaN nanorod LEDs from 5 to 300 K using a mode-locked Ti: sapphire laser (Coherent Mira 900, Santa Clara, CA, USA) with a wavelength of 800 nm as the optical excitation source. A third harmonic generator (THG, APE, Berlin, Germany) was used to excite the sample by an output wavelength of 267 nm. Laser power incident on the sample was kept below 1 mW and the sample was cooled by liquid nitrogen. From Figure 5a, it was obvious that with the increasing of the temperature, the PL peak at 325 nm, corresponding to the emission from AlGaN MQWs active region, did not significantly shift. The full width half maximum (FWHM) of the PL peak at 300 K was 11.57 nm. On the other hand, another minor peak at 334 nm could be observed at low temperature, possibly attributed to donor–acceptor pair (DAP) emission [25,26]. Figure 5b showed the temperature dependence of PL intensity of peaks at 325 and 334 nm. PL intensity had an apparent decrease with the increasing of the temperature, which was a characteristic behavior related to a thermally activated process. Figure 5c showed the normalized integrated PL intensity as a function of temperature. To estimate the internal quantum efficiency (IQE), most of the authors usually measured PL at a certain excitation condition and assume PL internal quantum efficiency at low temperature was equal to 100% regardless of nonradiative recombination [27]. The IQE could be approximately estimated by $I_{PL}(RT)/I_{PL}(LT)$, wherein $I_{PL}(RT)$ and $I_{PL}(LT)$ were the integrated PL intensity measured at room temperature (300 K) and low temperature (5 K), respectively [28,29]. The internal quantum efficiency of our sample at room temperature was calculated as 2.6%. Liu et al. [30] considered that except for the excitation power, the internal quantum efficiency of nanowire/nanorod devices may also be affected by the presence of surface states/defects, and the large bandgap AlGaN shell coverage could improve the IQE by reducing the effect of surface recombination on the quantum efficiency. On the other hand, the MQWs structure and the growth parameters in our samples needed to be optimized to further improve the IQE. Furthermore, we investigated the electroluminescence of AlGaN nanorod LEDs at room temperature. As shown in Figure 5d, the AlGaN nanorod LEDs emitted violet electroluminescence at 50 mA.

To give a further insight of the crystalline quality of our nitride nanorods, the cathodoluminescence (CL) properties were investigated with an electron beam acceleration voltage of 15 kV at room temperature. Figure 6a shows the CL spectrum measured from the region in Figure 6b. Like PL, the CL peak at approximately 318 nm, from AlGaN MQWs active region, was clearly observed. The FWHM of the CL peak was 26.86 nm. From the CL mapping image at the wavelength of 318 nm, as shown in Figure 6c, almost all the AlGaN nanorods showed clear luminescence. In particular, strong luminescence was observed from some isolated nanorods, which had a hexagonal c-axis-oriented shape. It was suggested that these nanorods had a higher crystallinity and lower density of grain boundary than those coalesced. The coalescence of nanorods was mainly considered from two aspects. One was the density of the AlGaN nucleation islands, which determined the density of the nanorods. The density of nucleation islands was mainly affected by Al composition; a higher Al composition would lead to a higher nucleation density. Another was the lateral growth rate of the nanorods, which contributed to the coalescence between the nanorods. Lateral growth was mainly affected by temperature, pressure and V/III ratio. For example, a low V/III ratio promoted the vertical growth of the nanorods. The mechanism of the coalescence in nanorods growth required more research to further improve the controllability and crystal quality of nanorods. Furthermore, a weak yellow luminescence band could be observed in Figure 6a. To research the yellow luminescence band, the CL mapping was measured at the wavelength of 500 nm, as shown in Figure 6d. There was no detective emitting luminescence at 500 nm, which showed the yellow luminescence band was negligible and revealed the high quality and low defects density of AlGaN nanorod LEDs.

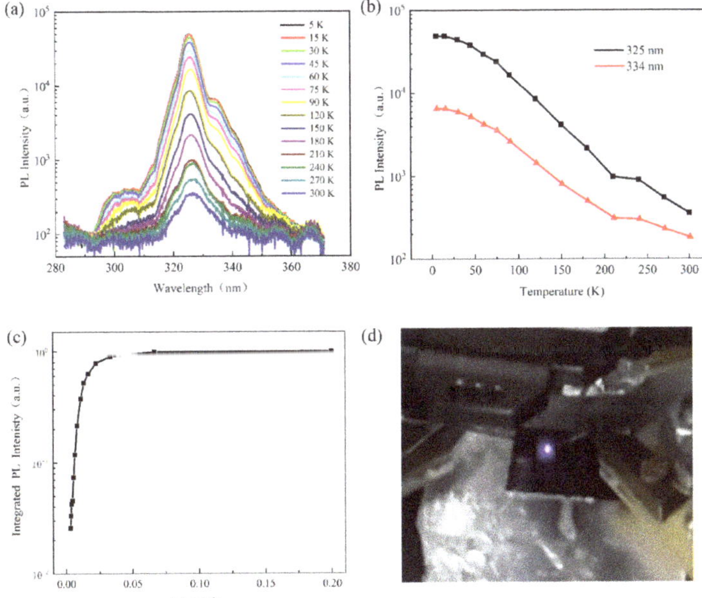


**Figure 5.** The photoluminescence properties and electroluminescence of AlGaN nanorod LEDs: (**a**) The spectra of temperature dependent PL; (**b**) the temperature dependence of PL intensity of peaks at 325 and 334 nm; (**c**) the temperature dependence of normalized integrated PL intensity of the AlGaN nanorod LEDs; and (**d**) optical image of the violet electroluminescence from the AlGaN nanorod LEDs.

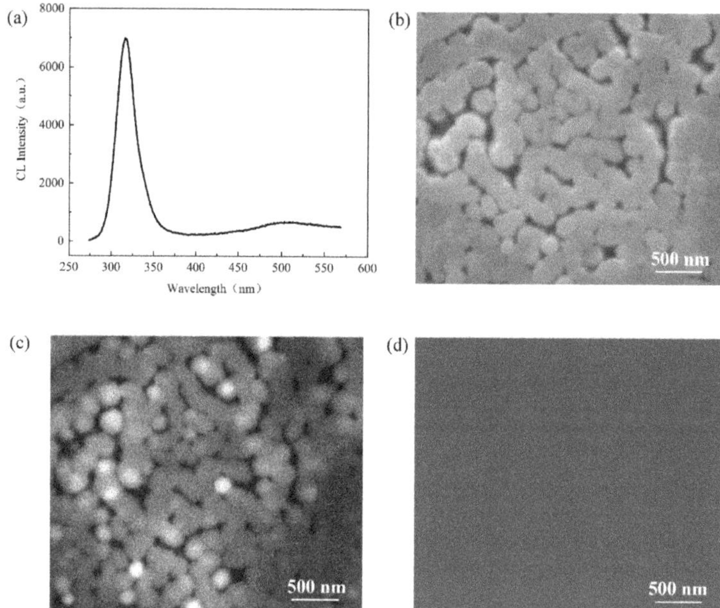


**Figure 6.** The cathodoluminescence properties of AlGaN nanorod LEDs: (**a**) The CL spectrum at room temperature; (**b**) the SEM image of CL mapping; (**c**) the CL mapping image at the wavelength of 318 nm; and (**d**) the CL mapping image at the wavelength of 500 nm.

## 4. Conclusions

In summary, we achieved the direct growth of AlGaN nanorod LEDs on single-layer graphene by metal organic chemical vapor deposition (MOCVD) without a metal catalyst. The nanorods were nucleated by AlGaN nucleation islands with a high Al composition, and included n-AlGaN, 6 period of AlGaN MQWs and p-AlGaN. The morphology, orientation, and crystal structure of the nanorods were characterized to show the uniform height and diameter as well as accordant orientation along the c-axis direction. Photoluminescence (PL) and cathodoluminescence (CL) were investigated to demonstrate the high optical properties and low defects density of AlGaN nanorods. The electroluminescence from the AlGaN nanorod LEDs was demonstrated. This method provides a novel way to grow nanorods without a metal catalyst or crystalline bulk. Furthermore, the growth of GaN nanorods on graphene buffer offers possibilities for the achievement of flexible optoelectronic devices.

**Author Contributions:** F.R. and Y.Y. contributed equally to this work. F.R., Y.Y., D.D., and H.W. conceived and designed the experiments. F.R., Y.Y., Y.W., and D.D. conducted the experiments. Z.L., G.Y., X.Y., and H.W. are responsible for technical assistance with nanorods fabrication and measurement. M.L., T.W., and J.Y. are responsible for technical assistance with graphene transfer. F.R., Y.W., and Z.L. performed the data analysis and wrote the manuscript. H.O., J.A., J.W., and J.L. contributed to the discussion and analysis of the results regarding the manuscript.

**Funding:** This research was funded by National Key R&D Program of China, Grant No. 2017YFB0403100, 2017YFB0403102; Beijing Municipal Science and Technology Project, Z161100002116032; Guangzhou Science & Technology Project of Guangdong Province, China, 201704030106 and 2016201604030035; Innovation Fund Denmark project, No. 4106-00018B.

**Conflicts of Interest:** The authors declare no conflict of interest.

## References


1. Zubia, D.; Hersee, S.D. Nanoheteroepitaxy: The Application of nanostructuring and substrate compliance to the heteroepitaxy of mismatched semiconductor materials. *J. Appl. Phys.* **1999**, *85*, 6492–6496. [CrossRef]
2. Li, S.; Waag, A. GaN-based nanorods for solid state lighting. *J. Appl. Phys.* **2012**, *111*, 5. [CrossRef]
3. Waag, A.; Wang, X.; Fündling, S.; Ledig, J.; Erenburg, M.; Neumann, R.; Al Suleiman, M.; Merzsch, S.; Wei, J.; Li, S.; et al. The nanorod approach: GaN nanoleds for solid state lighting. *Phys. Status Solidi C* **2011**, *8*, 2296–2301. [CrossRef]
4. Djavid, M.; Mi, Z. Enhancing the light extraction efficiency of AlGaN deep ultraviolet light emitting diodes by using nanowire structures. *Appl. Phys. Lett.* **2016**, *108*, 051102. [CrossRef]
5. Coulon, P.M.; Kusch, G.; Le Boulbar, E.D.; Chausse, P.; Bryce, C.; Martin, R.W.; Shields, P.A. Hybrid top-down/bottom-up fabrication of regular arrays of AlN nanorods for deep-UV core-shell leds. *Phys. Status Solidi B* **2018**, *255*, 1700445. [CrossRef]
6. Coulon, P.M.; Kusch, G.; Martin, R.W.; Shields, P.A. Deep UV emission from highly ordered AlGaN/AlN core-shell nanorods. *ACS Appl. Mater. Interfaces* **2018**, *10*, 33441–33449. [CrossRef] [PubMed]
7. Zhuang, Z.; Guo, X.; Zhang, G.; Liu, B.; Zhang, R.; Zhi, T.; Tao, T.; Ge, H.; Ren, F.; Xie, Z.; et al. Large-scale fabrication and luminescence properties of GaN nanostructures by a soft UV-curing nanoimprint lithography. *Nanotechnology* **2013**, *24*, 405303. [CrossRef] [PubMed]
8. Wu, S.; Wang, L.; Yi, X.; Liu, Z.; Wei, T.; Yuan, G.; Wang, J.; Li, J. Influence of lateral growth on the optical properties of GaN nanowires grown by hydride vapor phase epitaxy. *J. Appl. Phys.* **2017**, *122*, 205302. [CrossRef]
9. Wu, S.; Wang, L.; Yi, X.; Liu, Z.; Yan, J.; Yuan, G.; Wei, T.; Wang, J.; Li, J. Crystallographic orientation control and optical properties of GaN nanowires. *RSC Adv.* **2018**, *8*, 2181–2187. [CrossRef]
10. Li, Y.; Zhao, Y.; Wei, T.; Liu, Z.; Duan, R.; Wang, Y.; Zhang, X.; Wu, Q.; Yan, J.; Yi, X.; et al. Van Der Waals epitaxy of GaN-based light emitting diodes on wet-transferred multilayer graphene film. *Jpn. J. Appl. Phys.* **2017**, *56*, 085506. [CrossRef]
11. Mazid Munshi, A.; Weman, H. Advances in semiconductor nanowire growth on graphene. *Phys. Status Solidi Rapid Res. Lett.* **2013**, *7*, 713–726. [CrossRef]
12. Liu, X.; Wang, F.; Wu, H.; Wang, W. Strengthening metal nanolaminates under shock compression through dual effect of strong and weak graphene interface. *Appl. Phys. Lett.* **2014**, *104*, 231901. [CrossRef]
13. Kumaresan, V.; Largeau, L.; Madouri, A.; Glas, F.; Zhang, H.; Oehler, F.; Cavanna, A.; Babichev, A.; Travers, L.; Gogneau, N.; et al. Epitaxy of GaN nanowires on graphene. *Nano Lett.* **2016**, *16*, 4895–4902. [CrossRef] [PubMed]
14. Chung, K.; Yoo, H.; Hyun, J.K.; Oh, H.; Tchoe, Y.; Lee, K.; Baek, H.; Kim, M.; Yi, G.C. Flexible GaN light emitting diodes using GaN microdisks epitaxial laterally overgrown on graphene dots. *Adv. Mater.* **2016**, *28*, 7688–7694. [CrossRef] [PubMed]
15. Heilmann, M.; Munshi, A.M.; Sarau, G.; Gobelt, M.; Tessarek, C.; Fauske, V.T.; van Helvoort, A.T.; Yang, J.; Latzel, M.; Hoffmann, B.; et al. Vertically oriented growth of GaN nanorods on Si using graphene as an atomically thin buffer layer. *Nano Lett.* **2016**, *16*, 3524–3532. [CrossRef] [PubMed]
16. Zeng, Q.; Chen, Z.; Zhao, Y.; Wei, T.; Chen, X.; Zhang, Y.; Yuan, G.; Li, J. Graphene-assisted growth of high-quality AlN by metalorganic chemical vapor deposition. *Jpn. J. Appl. Phys.* **2016**, *55*, 085501. [CrossRef]
17. Qi, Y.; Wang, Y.; Pang, Z.; Dou, Z.; Wei, T.; Gao, P.; Zhang, S.; Xu, X.; Chang, Z.; Deng, B.; et al. Fast Growth of strain-free AlN on graphene-buffered sapphire. *J. Am. Chem. Soc.* **2018**, *140*, 11935–11941. [CrossRef] [PubMed]
18. Gupta, P.; Rahman, A.A.; Hatui, N.; Gokhale, M.R.; Deshmukh, M.M.; Bhattacharya, A. MOVPE growth of semipolar III-nitride semiconductors on CVD graphene. *J. Cryst. Growth* **2013**, *372*, 105–108. [CrossRef]
19. Lin, Y.T.; Yeh, T.W.; Dapkus, P.D. Mechanism of selective area growth of GaN nanorods by pulsed mode metalorganic chemical vapor deposition. *Nanotechnology* **2012**, *23*, 465601. [CrossRef] [PubMed]
20. Lin, Y.T.; Yeh, T.W.; Nakajima, Y.; Dapkus, P.D. Catalyst-free GaN nanorods synthesized by selective area growth. *Adv. Funct. Mater.* **2014**, *24*, 3162–3171. [CrossRef]
21. Journot, T.; Bouchiat, V.; Gayral, B.; Dijon, J.; Hyot, B. Self-assembled UV photodetector made by direct epitaxial GaN growth on graphene. *ACS Appl. Mater. Interfaces* **2018**, *10*, 18857–18862. [CrossRef] [PubMed]

22. Calizo, I.; Bejenari, I.; Rahman, M.; Liu, G.; Balandin, A.A. Ultraviolet Raman microscopy of single and multilayer graphene. *J. Appl. Phys.* **2009**, *106*, 043509. [CrossRef]
23. Sarau, G.; Lahiri, B.; Banzer, P.; Gupta, P.; Bhattacharya, A.; Vollmer, F.; Christiansen, S. Enhanced Raman scattering of graphene using arrays of split ring resonators. *Adv. Opt. Mater.* **2013**, *1*, 151–157. [CrossRef]
24. Zafar, Z.; Ni, Z.H.; Wu, X.; Shi, Z.X.; Nan, H.Y.; Bai, J.; Sun, L.T. Evolution of Raman spectra in nitrogen doped graphene. *Carbon* **2013**, *61*, 57–62. [CrossRef]
25. Chae, S.J.; Kim, Y.H.; Seo, T.H.; Duong, D.L.; Lee, S.M.; Park, M.H.; Kim, E.S.; Bae, J.J.; Lee, S.Y.; Jeong, H.; et al. Direct growth of etch pit-free GaN crystals on few-layer graphene. *RSC Adv.* **2015**, *5*, 1343–1349. [CrossRef]
26. Reshchikov, M.A.; Morkoç, H. Luminescence properties of defects in GaN. *J. Appl. Phys.* **2005**, *97*, 5–19. [CrossRef]
27. Watanabe, S.; Yamada, N.; Nagashima, M.; Ueki, Y.; Sasaki, C.; Yamada, Y.; Taguchi, T.; Tadatomo, K.; Okagawa, H.; Kudo, H. Internal quantum efficiency of highly-efficient $In_xGa_{1-x}N$-based near-ultraviolet light emitting diodes. *Appl. Phys. Lett.* **2003**, *83*, 4906–4908. [CrossRef]
28. Mickevičius, J.; Jurkevičius, J.; Kadys, A.; Tamulaitis, G.; Shur, M.; Shatalov, M.; Yang, J.; Gaska, R. Low-temperature redistribution of non-thermalized carriers and its effect on efficiency droop in AlGaN epilayers. *J. Phys. D Appl. Phys.* **2015**, *48*, 275105. [CrossRef]
29. Choi, J.K.; Huh, J.H.; Kim, S.D.; Moon, D.; Yoon, D.; Joo, K.; Kwak, J.; Chu, J.H.; Kim, S.Y.; Park, K.; et al. One-step graphene coating of heteroepitaxial GaN films. *Nanotechnology* **2012**, *23*, 435603. [CrossRef] [PubMed]
30. Liu, X.; Le, B.H.; Woo, S.Y.; Zhao, S.; Pofelski, A.; Botton, G.A.; Mi, Z. Selective area epitaxy of AlGaN nanowire arrays across nearly the entire compositional range for deep ultraviolet photonics. *Opt. Express* **2017**, *25*, 30494–30502. [CrossRef] [PubMed]

*Article*

# Direct van der Waals Epitaxy of Crack-Free AlN Thin Film on Epitaxial $WS_2$

**Yue Yin [1,2,3,†], Fang Ren [1,2,3,†], Yunyu Wang [1,2,3], Zhiqiang Liu [1,2,3,*], Jinping Ao [4], Meng Liang [1,2,3], Tongbo Wei [1,2,3], Guodong Yuan [1,2,3], Haiyan Ou [5], Jianchang Yan [1,2,3,*], Xiaoyan Yi [1,2,3,*], Junxi Wang [1,2,3] and Jinmin Li [1,2,3,*]**

1. Research and Development Center for Solid State Lighting, Institute of Semiconductors, Chinese Academy of Sciences, Beijing 100083, China; yinyue@semi.ac.cn (Y.Y.); rf@semi.ac.cn (F.R.); wyyu@semi.ac.cn (Y.W.); liangmeng@semi.ac.cn (M.L.); tbwei@semi.ac.cn (T.W.); gdyuan@semi.a.cn (G.Y.); jxwang@semi.ac.cn (J.W.)
2. Center of mAterials Science and Optoelectronics Engineering, University of Chinese Academy of Sciences, Beijing 100049, China
3. Beijing Engineering Research Center for the 3rd Generation Semiconductor mAterials and Application, Beijing 100083, China
4. Department of Electrical and Electronic Engineering, the University of Tokushima, 2-1, Minami-josanjima, Tokushima 770-8506, Japan; jpao@xidian.edu.cn
5. Department of Photonics Engineering, Technical University of Denmark, Ørsteds Plads 345A, DK-2800 Kongens Lyngby, Denmark; haou@fotonik.dtu.dk

* Correspondence: lzq@semi.ac.cn (Z.L.); yanjc@semi.ac.cn (J.Y.); spring@semi.ac.cn (X.Y.); jmli@semi.ac.cn (J.L.); Tel.: +86-010-8230-5423 (Z.L.)

† These authors contributed equally to this work.

Received: 9 November 2018; Accepted: 1 December 2018; Published: 4 December 2018

**Abstract:** Van der Waals epitaxy (vdWE) has drawn continuous attention, as it is unlimited by lattice-mismatch between epitaxial layers and substrates. Previous reports on the vdWE of III-nitride thin film were mAinly based on two-dimensional (2D) mAterials by plasma pretreatment or pre-doping of other hexagonal mAterials. However, it is still a huge challenge for single-crystalline thin film on 2D mAterials without any other extra treatment or interlayer. Here, we grew high-quality single-crystalline AlN thin film on sapphire substrate with an intrinsic $WS_2$ overlayer ($WS_2$/sapphire) by metal-organic chemical vapor deposition, which had surface roughness and defect density similar to that grown on conventional sapphire substrates. Moreover, an AlGaN-based deep ultraviolet light emitting diode structure on $WS_2$/sapphire was demonstrated. The electroluminescence (EL) performance exhibited strong emissions with a single peak at 283 nm. The wavelength of the single peak only showed a faint peak-position shift with increasing current to 80 mA, which further indicated the high quality and low stress of the AlN thin film. This work provides a promising solution for further deep-ultraviolet (DUV) light emitting electrodes (LEDs) development on 2D mAterials, as well as other unconventional substrates.

**Keywords:** AlN thin film; $WS_2$; MOCVD; van der Waals epitaxy

---

## 1. Introduction

Over the past few years, the van der Waals epitaxy (vdWE) of III-nitride devices has attracted a tremendous amount of attention [1–7]. This epitaxial mechanism is based on the weak van der Waals interaction between underlying two-dimensional (2D) mAterials and epitaxial layers, which will help to address the issue of lattice- and thermal-mismatch in the III-nitride heteroepitaxy [8,9]. Furthermore, semiconductors grown on 2D mAterials can be easily transferred to other unconventional substrates, which will create a new era for their potential applications in flexible electronics [10].

Among various 2D mAterials, graphene has been a focus due to the key advantage of its honeycomb crystal lattices, which are structurally compatible with the III-nitride film [7]. However, because of the lack of dangling bonds on 2D mAterials, the growth of high-quality III-nitride film is not an easy task. Several methods have been employed to create artificial defects, which are helpful to increase nucleation density for the subsequent growth of high-quality thin film [11]. Chung et al. conducted the growth of heteroepitaxial nitride thin film on high-density, vertically aligned ZnO nanowalls deposited on a graphene layer treated by $O_2$ plasma [1]. Han et al. utilized graphene oxide microscale patterns based on sapphire substrate to realize the epitaxial lateral overgrowth of GaN [2]. Kim et al. employed the periodic nucleation sites at the step edges of graphene/SiC to realize the direct vdWE of high-quality single-crystalline GaN film [3].

AlN is the fundamental component for AlGaN-based deep-UV LEDs, which are widely applied in the field of water purification, sensing, polymerization solidification, and non-line-of-sight communication [12]. Although some progress has been mAde in the growth of GaN thin films on 2D mAterials, the growth of AlN thin films remains challenging. Our group previously reported a series of studies on the growth of AlN thin films. Qing Zeng et al. released their research into the growth of continuous AlN film on graphene, with the step edges and defects as the nucleation sites [5]. Yang Li et al. experimentally studied the feasibility of solving large mismatch problems with multilayer graphene acting as the interlayer between sapphire and the III-nitride, and further studied the effects of the optical and electrical properties of LEDs on graphene [6]. To mAke the action in strict van der Waals epitaxial growth on 2D mAterials interlayer clear, Yunyu Wang et al. investigated the roles of a graphene buffer layer in AlN nucleation on a sapphire substrate, indicating that graphene caused a decrease of nucleation density and an increase in AlN nuclei growth rate, and significantly weakened the AlN–$Al_2O_3$ interaction to release the strain [7].

The studies mentioned above are all based on the graphene buffer layer. Auxiliary methods were needed to assist the deposition of the III-nitride film (e.g., plasma treatment and ZnO nanowalls) [1,3,4]. However, the growth mechanism was changed owing to the introduction of dangling bonds, which means it is not a van der Waals epitaxy in the true sense. To realize the strict van der Waals epitaxy, more 2D mAterials are tested for the growth of III-nitride thin film. $WS_2$ and $MoS_2$ would be perfect candidates because their small lattice mismatches with III-nitrides are only 1.0% and 0.8%, respectively, to the "a" lattice parameter of GaN [13]. In 2016, Gupta et al. proposed exhaustive studies on the growth of strain-free and single-crystal GaN islands by metal-organic chemical vapor deposition (MOCVD) on mechanically-exfoliated $WS_2$ flakes [13]. Chao Zhao et al. reported the growth of InGaN/GaN nanowire LEDs on sulfurized Mo substrates [14]. Nevertheless, the growth of continuous III-nitride thin film on transition metal dichalcogenide (TMDC) buffer layers still lacks relevant research.

Motivated by these considerations, here we present the first experimental investigation of the direct epitaxy of continuous AlN thin film with a $WS_2$ interlayer. Herein, high-quality AlN was obtained by MOCVD on intrinsic $WS_2$/sapphire substrate. The measured root mean square (RMS) roughness was 0.230 nm. Thanks to the atomistic smoothness of the released AlN film, a fully functional 283 nm deep-ultraviolet (DUV) light emitting diodes (LEDs) device was further demonstrated.

## 2. Materials and Methods

In our work, high-purity $WO_3$ (at 1000 °C) and S (at 200 °C) were applied for the synthesis of single-crystalline $WS_2$ film with an area of $1 \times 1$ cm$^2$ on sapphire substrates directly, with Ar and $H_2$ as carrier gases, respectively, in a three-temperature zone tube furnace. An AlN thin film was deposited by MOCVD on the $WS_2$/sapphire substrate. Trimethylaluminum (TMAl) and $NH_3$ were employed as Al and N precursors, respectively. An AlN nucleation layer was first deposited at 890 °C for 4 min with a V/III ratio of about 9640. After low-temperature growth of the AlN nucleation layer, the temperature was increased to 1200 °C to grow a 500 nm AlN epilayer for 30 min with a V/III ratio of 578. No additional intermediate layers or substrate treatments were employed for AlN layer growth on $WS_2$/sapphire layer. $H_2$ was used as carrier gas for all of the growth steps. The MOCVD chamber pressure was kept at 50 torr during the whole growth process.

After the AlN thin film epitaxial growth, AlGaN-based DUV LED structures were further grown on the AlN/$WS_2$/Sapphire template. Trimethylgallium (TMGa) was used as the Ga precursor. A 20-period AlN/$Al_{0.6}Ga_{0.4}N$ superlattice (SL) was first deposited at 1130 °C, with periodic flow change of TMAl to adjust the deposition component, while the TMGa flow was kept at 32 sccm. Temperature was reduced to 1002 °C. Then, the $SiH_4$ lane was opened with the flow of 20 sccm, and an n-$Al_{0.6}Ga_{0.4}N$ layer was deposited with the thickness of 1.8 μm. Five-period $Al_{0.5}Ga_{0.5}N$/$Al_{0.6}Ga_{0.4}N$ multiple quantum wells (MQWs) were further grown, with a 12.2 nm quantum barrier and 2.4 nm quantum well in each period. The TMAl was switched from 24 sccm to 14 sccm, while the TMGa was switched from 8 sccm to 7 sccm alternatively each time before the growth of quantum wells. A 60 nm p-$Al_{0.65}Ga_{0.35}N$ electron blocking layer (EBL), a p-AlGaN cladding layer, and a p-GaN contact layer were subsequently extended. The $NH_3$ was 2500 sccm during the whole growth process. After the growth, the sample were annealed at 800 °C with $N_2$ flow for 20 min to activate the Mg acceptors.

Furthermore, standard LED processes were mAde to fabricate DUV LED, such as photolithography, ICP etching, etc. A 210 nm Ti/Al/Ti/Au metal stack and a 40 nm Ni/Au stack were respectively vapored as the n- and p-electrodes. In the end, the chips were flip-chip bonded onto ceramic submounts.

### 3. Results

The surface morphology of the $WS_2$ on the sapphire substrate was examined using a Hitachi S4800 scanning electron microscopy (SEM, Tokyo, Japan) operated at 3 keV acceleration voltage (Figure 1a) and tapping mode atomic force microscopy (AFM, D3100, Veeco, New York, NY, USA) (Figure 1b), indicating that the substrate could be fully covered with continuous and uniform monolayer $WS_2$ film. Some secondary nuclei were attached to the $WS_2$ film. A JOBIN YVON-HORIBA HR800 Raman spectrometer (Kyoto, Japan) with a semiconductor laser at 532 nm as the excitation source was employed to analyze the chemical properties and detailed compositions of the direct-grown $WS_2$ film. Raman spectra presented similar intensity of 2LA and $A_{1g}$ mode peaks (Figure 1c), which verified the good film uniformity over the area of $1 \times 1$ cm$^2$ [13].

After the AlN thin film was grown on the $WS_2$/sapphire substrate by MOCVD, a SEM image was obtained to investigate the surface morphology of the as-grown film, as presented in Figure 2a. Mirror-smooth and crack-free AlN thin films were grown with complete coalescence. The AFM image further verifies that the surface topology of as-grown AlN thin film on $WS_2$/sapphire substrates was flat, with the RMS roughness at 0.230 nm over a lateral distance of 5 μm, as seen in Figure 2b, which was comparable with the AlN thin film directly grown on sapphire [15].

The stress of as-grown AlN thin film was further evaluated by Raman spectroscopy. The biaxial strain in the AlN layer was relative to the $E_2$ phonon mode movement of the Raman spectrum. Figure 2c shows the AlN epilayers grown on $WS_2$/sapphire substrate sustained tensile stress, demonstrating a smaller frequency (656 cm$^{-1}$) compared with stress-free AlN (657 cm$^{-1}$). The stress relaxation of AlN epilayers, prompted by the $WS_2$ interlayer, can be appraised in light of $\Delta\omega = K\sigma_{xx}$. In this formula, $\Delta\omega$ is the $E_2$ peak movement between the sample and stress-free AlN crystal, while $K$ is the biaxial stress conversion factor ≈3.7 cm$^{-1}$·GPa$^{-1}$ [16–18]. The biaxial stress value of AlN epilayers grown on $WS_2$/sapphire substrate was 0.27 GPa. Compared with the Raman spectra of the $WS_2$/sapphire substrate, the presence of $WS_2$ after the film's growth was confirmed by the same peak at 417.6 cm$^{-1}$ in Figure 1c.

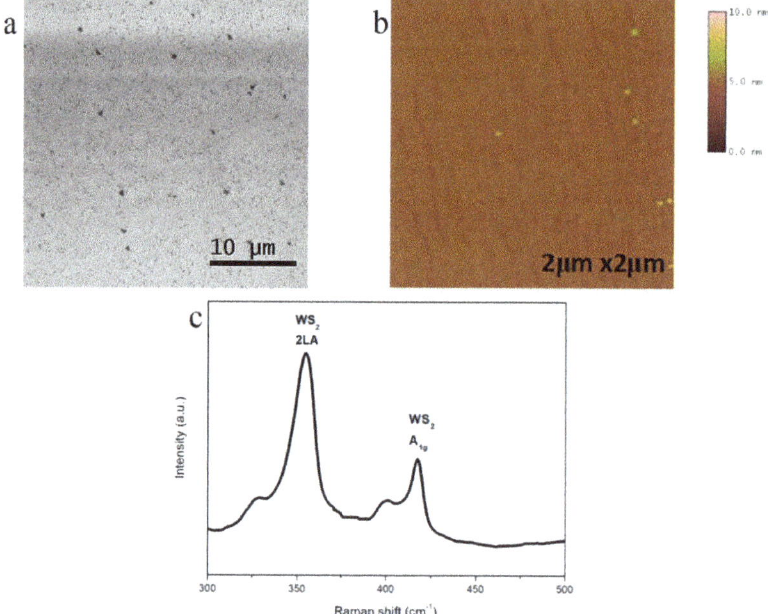


**Figure 1.** Characterizations of direct growth of $WS_2$ on sapphire substrate. (**a**) Scanning electron microscopy (SEM) image of the $WS_2$ film directly grown on sapphire substrates. (**b**) Atomic force microscopy (AFM) image of the $WS_2$ film with root mean square (RMS) roughness around 0.203 nm. (**c**) Raman spectra of the $WS_2$ film on sapphire substrates.

The crystal quality of AlN was evaluated by means of a Bede X-ray metrology double crystal high-resolution X-ray diffraction rocking curve (XRC) analyses. Figure 2d,e shows the ω-scan profiles (rocking curves) of the AlN (0002) and (10-12) peaks. The full width at half mAximum (FWHM) values of the (0002) and (10-12) rocking curve of AlN are directly related to the densities of screw- and edge-dislocations in epilayers. The FWHM values of AlN thin film were measured to be 546 arcsec and 1469 arcsec, respectively, for (0002) and (10-12) reflections. The estimated densities of screw and edge dislocations were $6.49 \times 10^8$ cm$^{-2}$ and $2.42 \times 10^{10}$ cm$^{-2}$ [19]. Although the FWHM value was slightly larger than that of the AlN thin film grown on graphene film with extra plasma treatment. The results of rocking curves suggest the preferable c-axis alignment of the AlN film grown on the $WS_2$ interlayer. To explore the epitaxial relationship between AlN epilayers and c-plane sapphire, we employed XRD-φ scan with $2\theta = 25.58°$ $\chi = 57.61°$ (Figure 2f). Six peaks of the AlN curve could be observed. Each one was 60 degrees apart, while three peaks of the sapphire curve could be observed, and each one was 120 degrees apart. Those curves reveal that the AlN (0002) facets were rotated by 30 degrees with sapphire (0006) facets through $WS_2$, describing the epitaxial relationship was [1100] AlN//[11-20] sapphire. The crystalline orientations of as-grown AlN film were also identified by using electron backscatter diffraction (EBSD). The EBSD mApping provided evidence that most of the area of the AlN film displayed almost (0001) single crystallinity, as demonstrated in red by the inverse pole figure color triangle (Figure 2g). These results all strongly suggest that single-crystalline AlN film was grown on $WS_2$ film, and the DUV LEDs could be subsequently deposited on the AlN/$WS_2$/sapphire template.

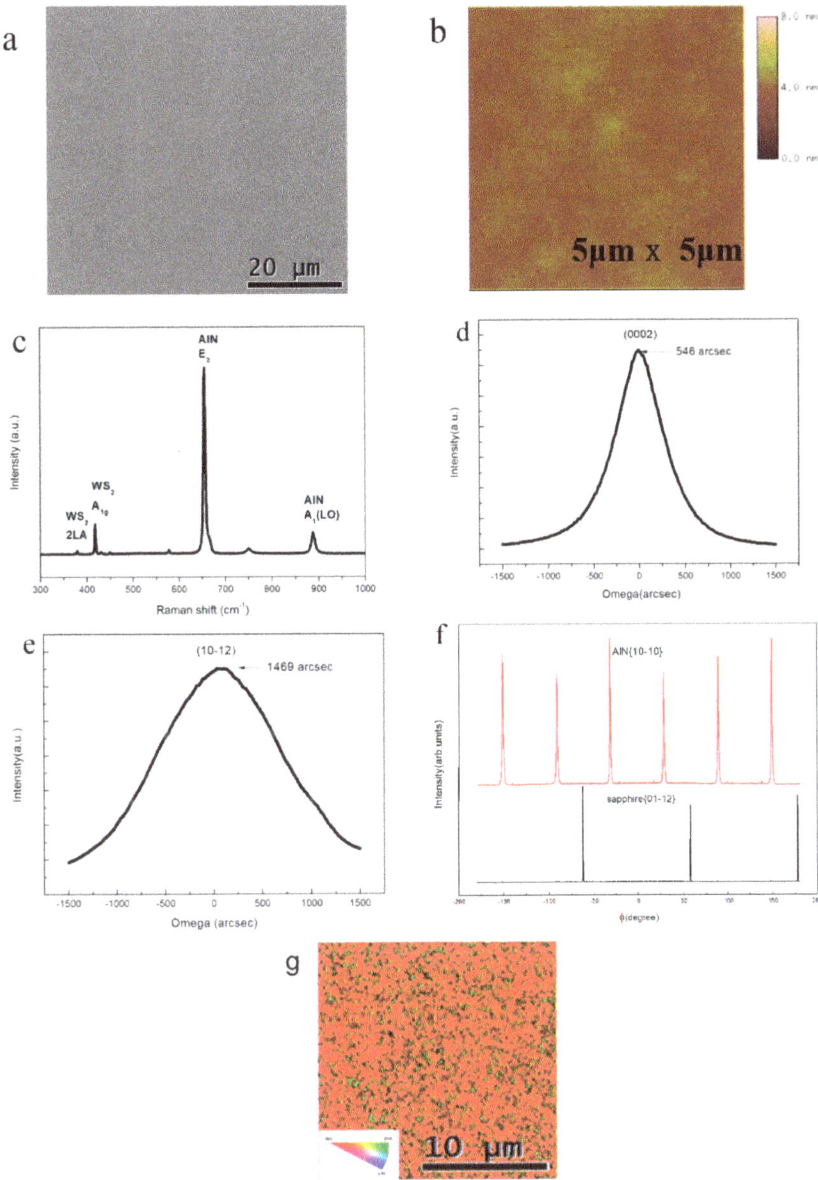


**Figure 2.** Characterizations of AlN thin film growth on $WS_2$/sapphire substrate without extra treatment. (**a**) SEM image, (**b**) AFM image, (**c**) Raman spectra, (**d**) X-ray rocking curves of (0002), and (**e**) (10-12) of the AlN film grown on sapphire with $WS_2$ interlayers. (**f**) X-ray powder diffraction (XRD) φ scan curve with 2θ = 25.58° χ = 57.61°. (**g**) Electron backscatter diffraction (EBSD) mApping of AlN film.

A conventional AlGaN-based DUV LED structure on $WS_2$/sapphire substrate was achieved after the growth of AlN thin film. Its schematic illustration is shown in Figure 3a. In order to characterize the LED heterojunction structure and confirm the existence of $WS_2$ in the AlN/$WS_2$/sapphire interface, cross-sectional scanning transmission electron microscopy (STEM) and energy dispersive X-ray spectroscopy (EDX) were applied. Microstructural behaviors of the whole heterojunction grown on AlN/$WS_2$/sapphire template were investigated using the cross-sectional STEM at low mAgnification, allowing us to scan the entire DUV LED microstructure. The cross-sectional STEM image in Figure 3b shows that layer-by-layer grown LED structures were formed, consistent with the schematic illustration. Figure 3c indicates the high quality of multiple quantum wells (MQWs) at a higher mAgnification, verifying that the five-period $Al_{0.5}Ga_{0.5}N/Al_{0.6}Ga_{0.4}N$ MQWs were defect-free. Figure 3d proves the existence of $WS_2$ after the growth of the LED structure. The atomically resolved STEM image shows clearly distinguishable line between AlN and sapphire as the signal of $WS_2$ exists. We also investigated the existence of $WS_2$ interlayers by using EDX. The elemental mApping confirmed the existence of $WS_2$ interlayers with distributions of S (Figure 3e) and W (Figure 3f). W element distribution was mAinly localized at the interface, with a relatively clear boundary. However, the wide distribution of S was probably the result of the decomposition of the $WS_2$ layer to some extent. We tend to believe that $WS_2$ layer still existed, although with mAny defects (e.g., S vacancy).

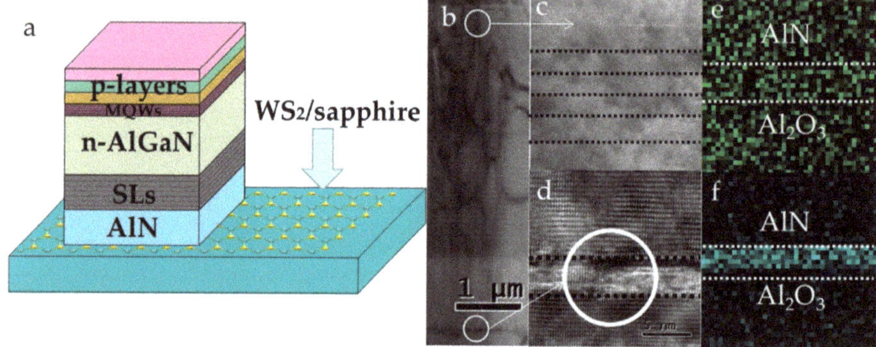


**Figure 3.** Characterizations of conventional AlGaN-based deep ultraviolet (DUV) light emitting diodes (LEDs) grown on $WS_2$/sapphire substrates. (**a**) Schematic illustration of the DUV LED structure. (**b**) Cross-sectional scanning transmission electron microscopy (STEM) image of heterojunction LEDs; (**c**) $Al_{0.5}Ga_{0.5}N/Al_{0.6}Ga_{0.4}N$ MQWs; and (**d**) the AlN/$WS_2$/sapphire interface of the as-grown DUV LED. (**e**) Energy dispersive X-Ray spectroscopy (EDX) mApping of S; and (**f**) W element showing the $WS_2$ gap between AlN and sapphire.

After the electrode deposition and other fabrication processes, the on-wafer electroluminescence (EL) performance of the DUV LED structure on the $WS_2$/sapphire substrates was further investigated. A single-peaked spectrum was observed, with a peak wavelength at 283 nm at a dividing current of 80 mA (Figure 4a). Moreover, the current-voltage curve of the DUV LED with $WS_2$ showed good rectifying behavior with a turn-on voltage of 3.38 V (Figure 4b), and the leakage current measured at −4 V was about 0.04 mA. This confirms that the quality of AlN thin film on $WS_2$/sapphire was sufficiently robust to fabricate DUV LEDs. Figure 4c shows the functional relationship between light-out power (LOP) and injection current of LEDs. LOP increased simultaneously with the injection current, revealing that the EL emission was generated from the carrier injection and radiative recombination at MQW layers. In order to evaluate the reliability, the normalized EL of as-fabricated LED under different injection currents were investigated as shown in Figure 4d. The wavelength of the single peak only showed faint peak-position redshift from 281.8 to 283 nm, with current increasing from 30 to 70 mA, then blueshifted to 282.6 nm under the injection current of 80 mA. The inevitable thermal effect of UV devices and threading dislocation caused the redshift, while screening of the polarization electric field in strained MQW structures caused the blueshift [20,21]. The faint peak-position shift should be attributed to the low-stress property of AlN thin film. The carrier's recombination in p-AlGaN cladding layer likely led to the weak shoulder at 324 nm with low injection current in the EL spectrum, which signifies that further optimization of the electron blocking layer to enhance the quantum confinement of electrons and suppress the electron overflow is necessary [22,23]. With increasing current, the relative intensity of the weak shoulder got weaker, until the weak shoulder vanished. These results demonstrate that conventional DUV LEDs could be fabricated on the $WS_2$/sapphire substrate.

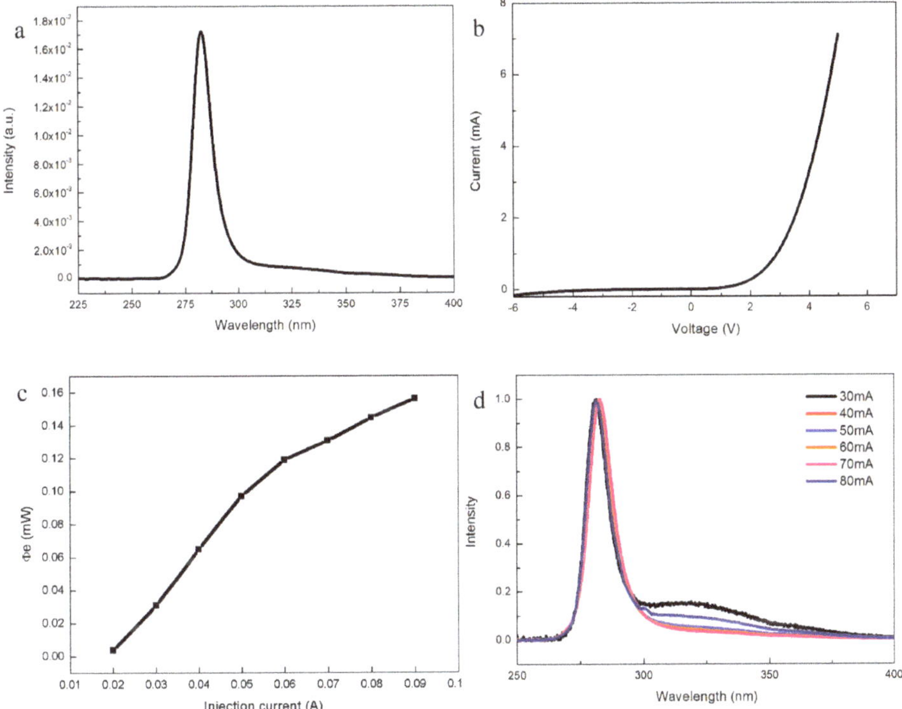


**Figure 4.** Electroluminescence (EL) of as-fabricated DUV LEDs. (**a**) The single-peaked EL spectrum of the DUV LED structure. (**b**) I-V curve of the fabricated DUV LEDs with $WS_2$ buffer layer. (**c**) Light-out power (LOP) of the fabricated LEDs at various injection currents. (**d**) The normalized EL spectra of fabricated LEDs with currents ranging from 30 to 80 mA.

## 4. Conclusions

We demonstrated the experimental realization of crack-free and mirror-like single-crystalline AlN thin film on $WS_2$ buffered sapphire substrate, resulting in RMS surface roughness of 0.230 nm, which is within the range of directly grown AlN film grown on sapphire substrate using MOCVD. The estimated densities of screw and edge dislocations were $6.49 \times 10^8$ cm$^{-2}$ and $2.42 \times 10^{10}$ cm$^{-2}$. Hence, the quality of AlN thin film on $WS_2$/sapphire was robust enough to fabricate DUV LEDs. Fully functional DUV LED was subsequently fabricated on the AlN/$WS_2$/sapphire template. Its clear EL emissions had a peak wavelength of 283 nm at 80 mA. The wavelength of the single peak only showed a faint peak-position shift with increasing current to 80 mA. The cross-sectional TEM and EDS results confirmed our growth model and the presence of the continuous $WS_2$ layer in the AlN/$WS_2$/sapphire hetero-interface, even after the growth of LED. The efficient DUV LEDs fabricated on $WS_2$/sapphire show the potential of $WS_2$ for the epitaxy of the III-nitride on large-size and low-cost metal or amorphous substrates in the future. Our work provides a potential solution for further DUV LED development on unconventional substrates.

**Author Contributions:** Y.Y., F.R., Y.W., Z.L., J.A., M.L., J.Y., and J.L. conceived and designed the experiments. Y.Y. and F.R. conducted the experiments. Y.W., Z.L., T.W., G.Y., M.L., X.Y., H.O., and J.Y. are responsible for technical assistance with LED fabrication and measurement. Y.Y. performed the data analysis and wrote the mAnuscript. All authors contributed to the discussion and analysis of the results regarding the manuscript.

**Funding:** This research was funded by National Key R&D Program of China, grant number 2017YFB0403100, 2017YFB0403103, Beijing Municipal Science and Technology Project, grant number Z161100002116032, Guangzhou Science & Technology Project of Guangdong Province, China, grant number 201704030106 and 2016201604030035 and Innovation Fund Denmark, grant number 4106-00018B.

**Conflicts of Interest:** The authors declare no conflict of interest.

## References

1. Chung, K.; Lee, C.H.; Yi, G.C. Transferable GaN layers grown on ZnO-coated graphene layers for optoelectronic devices. *Science* **2010**, *330*, 655–657. [CrossRef]
2. Han, N.; Cuong, T.V.; Han, M.; Ryu, B.D.; Chandramohan, S.; Park, J.B.; Kang, J.H.; Park, Y.J.; Ko, K.B.; Kim, H.Y.; et al. Improved heat dissipation in gallium nitride light-emitting diodes with embedded graphene oxide pattern. *Nat. Commun.* **2013**, *4*, 1452. [CrossRef]
3. Kim, J.; Bayram, C.; Park, H.; Cheng, C.W.; Dimitrakopoulos, C.; Ott, J.A.; Reuter, K.B.; Bedell, S.W.; Sadana, D.K. Principle of direct van der Waals epitaxy of single-crystalline films on epitaxial graphene. *Nat. Commun.* **2014**, *5*, 4836. [CrossRef] [PubMed]
4. Nepal, N.; Wheeler, V.D.; Anderson, T.J.; Kub, F.J.; mAstro, M.A.; Myers-Ward, R.L.; Qadri, S.B.; Freitas, J.A.; Hernandez, S.C.; Nyakiti, L.O.; et al. Epitaxial growth of III-Nitride/graphene heterostructures for electronic devices. *Appl. Phys. Express* **2013**, *6*, 061003. [CrossRef]
5. Zeng, Q.; Chen, Z.; Zhao, Y.; Wei, T.; Chen, X.; Zhang, Y.; Yuan, G.; Li, J. Graphene-assisted growth of high-quality AlN by metalorganic chemical vapor deposition. *Jpn. J. Appl. Phys.* **2016**, *55*, 085501. [CrossRef]
6. Li, Y.; Zhao, Y.; Wei, T.; Liu, Z.; Duan, R.; Wang, Y.; Zhang, X.; Wu, Q.; Yan, J.; Yi, X.; et al. Van der Waals epitaxy of GaN-based light-emitting diodes on wet-transferred multilayer graphene film. *Jpn. J. Appl. Phys.* **2017**, *56*, 085506. [CrossRef]
7. Qi, Y.; Wang, Y.; Pang, Z.; Dou, Z.; Wei, T.; Gao, P.; Zhang, S.; Xu, X.; Chang, Z.; Deng, B.; et al. Fast Growth of Strain-Free AlN on Graphene-Buffered Sapphire. *J. Am. Chem. Soc.* **2018**, *140*, 11935–11941. [CrossRef] [PubMed]
8. Utama, M.I.; Zhang, Q.; Zhang, J.; Yuan, Y.; Belarre, F.J.; Arbiol, J.; Xiong, Q. Recent developments and future directions in the growth of nanostructures by van der Waals epitaxy. *Nanoscale* **2013**, *5*, 3570–3588. [CrossRef]
9. Koma, A.; Sunouchi, K.; Miyajima, T. Fabrication and characterization of heterostructures with subnanometer thickness. *Microelectron. Eng.* **1984**, *2*, 129–136. [CrossRef]
10. Das, T.; Sharma, B.K.; Katiyar, A.K.; Ahn, J.-H. Graphene-based flexible and wearable electronics. *J. Semicond.* **2018**, *39*, 011007. [CrossRef]
11. Choi, J.H.; Kim, J.; Yoo, H.; Liu, J.; Kim, S.; Baik, C.-W.; Cho, C.-R.; Kang, J.G.; Kim, M.; Braun, P.V.; et al. Heteroepitaxial Growth of GaN on Unconventional Templates and Layer-Transfer Techniques for Large-Area, Flexible/Stretchable Light-Emitting Diodes. *Adv. Opt. mAter.* **2016**, *4*, 505–521. [CrossRef]
12. Li, J.; Liu, Z.; Liu, Z.; Yan, J.; Wei, T.; Yi, X.; Wang, J. Advances and prospects in nitrides based light-emitting-diodes. *J. Semicond.* **2016**, *37*, 061001.
13. Gupta, P.; Rahman, A.A.; Subramanian, S.; Gupta, S.; Thamizhavel, A.; Orlova, T.; Rouvimov, S.; Vishwanath, S.; Protasenko, V.; Laskar, M.R.; et al. Layered transition metal dichalcogenides: Promising near-lattice-matched substrates for GaN growth. *Sci. Rep.* **2016**, *6*, 23708. [CrossRef] [PubMed]
14. Zhao, C.; Ng, T.K.; Tseng, C.C.; Li, J.; Shi, Y.M.; Wei, N.N.; Zhang, D.L.; Consiglio, G.B.; Prabaswara, A.; Alhamoud, A.A.; et al. InGaN/GaN nanowires epitaxy on large-area $MoS_2$ for high-performance light-emitters. *RSC Adv.* **2017**, *7*, 26665–26672. [CrossRef]
15. Imura, M.; Nakano, K.; Kitano, T.; Fujimoto, N.; Narita, G.; Okada, N.; Balakrishnan, K.; Iwaya, M.; Kamiyama, S.; Amano, H.; et al. Microstructure of epitaxial lateral overgrown AlN on trench-patterned AlN template by high-temperature metal-organic vapor phase epitaxy. *Appl. Phys. Lett.* **2006**, *89*, 221901. [CrossRef]
16. Zhao, D.G.; Xu, S.J.; Xie, M.H.; Tong, S.Y.; Yang, H. Stress and its effect on optical properties of GaN epilayers grown on Si(111), 6H-SiC(0001), and c-plane sapphire. *Appl. Phys. Lett.* **2003**, *83*, 677–679. [CrossRef]

17. Park, A.H.; Seo, T.H.; Chandramohan, S.; Lee, G.H.; Min, K.H.; Lee, S.; Kim, M.J.; Hwang, Y.G.; Suh, E.K. Efficient stress-relaxation in InGaN/GaN light-emitting diodes using carbon nanotubes. *Nanoscale* **2015**, *7*, 15099–15105. [CrossRef] [PubMed]
18. Qi, L.; Xu, Y.; Li, Z.; Zhao, E.; Yang, S.; Cao, B.; Zhang, J.; Wang, J.; Xu, K. Stress analysis of transferable crack-free gallium nitride microrods grown on graphene/SiC substrate. *Mater. Lett.* **2016**, *185*, 315–318. [CrossRef]
19. Srikant, V.; Speck, J.S.; Clarke, D.R. Mosaic structure in epitaxial thin films having large lattice mismatch. *J. Appl. Phys.* **1997**, *82*, 4286–4295. [CrossRef]
20. Dong, P.; Yan, J.; Zhang, Y.; Wang, J.; Zeng, J.; Geng, C.; Cong, P.; Sun, L.; Wei, T.; Zhao, L.; et al. AlGaN-based deep ultraviolet light-emitting diodes grown on nano-patterned sapphire substrates with significant improvement in internal quantum efficiency. *J. Cryst. Growth* **2014**, *395*, 9–13. [CrossRef]
21. Kuokstis, E.; Yang, J.W.; Simin, G.; Khan, M.A.; Gaska, R.; Shur, M.S. Two mechanisms of blueshift of edge emission in InGaN-based epilayers and multiple quantum wells. *Appl. Phys. Lett.* **2002**, *80*, 977–979. [CrossRef]
22. Yan, J.; Wang, J.; Zhang, Y.; Cong, P.; Sun, L.; Tian, Y.; Zhao, C.; Li, J. AlGaN-based deep-ultraviolet light-emitting diodes grown on high-quality AlN template using MOVPE. *J. Cryst. Growth* **2015**, *414*, 254–257. [CrossRef]
23. Shatalov, M.; Sun, W.; Lunev, A.; Hu, X.; Dobrinsky, A.; Bilenko, Y.; Yang, J.; Shur, M.; Gaska, R.; Moe, C.; et al. AlGaN deep-ultraviolet light-emitting diodes with external quantum efficiency above 10%. *Appl. Phys. Express* **2012**, *5*, 082101. [CrossRef]

Article

# Properties of Undoped Few-Layer Graphene-Based Transparent Heaters

**Yong Zhang [1,2,*], Hao Liu [1], Longwang Tan [1], Yan Zhang [1], Kjell Jeppson [2], Bin Wei [3] and Johan Liu [1,2,*]**

1. SMIT Center, School of Mechatronic Engineering and Automation, Shanghai University, Changzhong Road, Shanghai 201800, China; LiuHao_T@shu.edu.cn (H.L.); LonggoingTan@i.shu.edu.cn (L.T.); yzhang@staff.shu.edu.cn (Y.Z.)
2. Electronics Materials and Systems Laboratory, Department of Microtechnology and Nanoscience, Chalmers University of Technology, SE-41296 Gothenburg, Sweden; kjell.jeppson@chalmers.se
3. Key Laboratory of Advanced Display and System Applications, Ministry of Education, Shanghai University, Yanchang Road, Shanghai 200072, China; bwei@shu.edu.cn

* Correspondence: yongz@shu.edu.cn (Y.Z.); johan.liu@chalmers.se (J.L.)

Received: 14 November 2019; Accepted: 17 December 2019; Published: 24 December 2019

**Abstract:** In many applications like sensors, displays, and defoggers, there is a need for transparent and efficient heater elements produced at low cost. For this reason, we evaluated the performance of graphene-based heaters with from one to five layers of graphene on flexible and transparent polyethylene terephthalate (PET) substrates in terms of their electrothermal properties like heating/cooling rates and steady-state temperatures as a function of the input power density. We found that the heating/cooling rates followed an exponential time dependence with a time constant of just below 6 s for monolayer heaters. From the relationship between the steady-state temperatures and the input power density, a convective heat-transfer coefficient of 60 $W{\cdot}m^{-2}{\cdot}°C^{-1}$ was found, indicating a performance much better than that of many other types of heaters like metal thin-film-based heaters and carbon nanotube-based heaters.

**Keywords:** graphene; chemical vapor deposition (CVD); transfer; heater; resistance; heating/cooling rates

## 1. Introduction

Transparent resistive heaters were proposed for a variety of applications, such as sensors [1], displays [2], defoggers [3], and defrosters [4]. For certain applications, films of indium tin oxide (ITO) are commonly used materials for transparent heaters; however, poor stretchability and a complicated and costly fabrication process limit their usage [5]. Much effort was devoted to developing replacement materials, with some examples being silver nanowires [6–8], carbon nanotube films [9,10], and hybrid composites [11].

The electrothermal properties of graphene, an atom-thick planar sheet of $sp^2$-bonded carbon atoms in a honeycomb pattern [12–14], with superconductivity recently observed in magic-angle graphene superlattices [15], indicate that this two-dimensional (2D) material could be the perfect material for many applications including transparent heater applications. Consequently, graphene-based heaters were recently proposed with graphene obtained by chemical vapor deposition (CVD) [3], from reduced graphene oxide [16–18], and from graphene aerogels [19]. Among these methods, CVD appears to be the most attractive for industrial production of graphene because of its scalability [20]. Graphene-based heaters fabricated by CVD are often doped with $AuCl_3$, $Au\text{-}CH_3NO_2$, or $HNO_3$ to enhance their electrothermal performance [21]. However, dopants introduced into graphene films might affect the stability of the material by reacting with ambient molecules, thereby causing material properties to degrade over time.

In this article, we present the results of a study of the electrothermal properties of transparent undoped few-layer graphene-based heaters where from one to five layers of graphene grown by CVD were transferred to flexible polyethylene terephthalate (PET) substrates. Our observations of their heating/cooling rates at different power densities are reported in the upcoming sections.

## 2. Fabrication and Evaluation of Graphene-Based Heater Samples

For the set of experiments presented in this work, monolayer graphene was grown on copper foils under low partial pressure by CVD following a standard procedure previously described in detail [22]. In short, this process involves a ~15-min temperature ramp-up to 1000 °C in ambient argon (1000 sccm)/hydrogen (80 sccm), a 5-min annealing at the growth temperature, a 5-min growth period using a methane precursor (5 sccm), and a ~35-min cool-down to room temperature (RT) again in ambient argon/hydrogen.

After growth, a standard layer-by-layer approach was employed to transfer the graphene from the copper foil onto PET substrates involving spin-coated poly(methyl methacrylate) (PMMA). By repeating the transfer process, a set of samples with between one and five layers of graphene was obtained. Samples were then turned into heaters by deposition of Cr/Au electrodes with a thickness of 10/70 nm at the edges of the graphene on PET samples. Finally, T-type thermocouples were attached to the back side of the PET substrates for in situ monitoring of the heater temperature by using a Keysight 34,970 A data acquisition/data logger switch unit. The accuracy of the thermocouple was estimated to be ±0.5 °C. All measurements were performed in a fume cupboard.

After fabrication, the quality of the graphene was investigated. Optical images showing the morphology of the Cu foil after graphene growth are shown in Figure 1a. As seen using an optical microscope, grain boundaries up to several hundred micrometers long became visible on the Cu foil. These grain boundaries were more clearly identified in scanning electron microscope (SEM) photos, such as the one shown in Figure 1b. The wrinkles were due to the mismatch between the coefficients of thermal expansion of graphene and the underlying metal [23]. It should be noted that those wrinkles crossed the Cu grain boundaries, and that no islands were observed, indicating that the as-grown graphene film was continuous [24]. From the Raman spectrum (Raman measurements were carried out with an XploRA (HORIBA, Ltd. Kyoto, Japan) at a 638-nm excitation wavelength and a 100× objective, with an incident power of ~1 mW) in Figure 1c, typical 2D/G peak intensity ratios of ~2.9 were identified indicating monolayer graphene [25]. As no D band was observed, the Raman spectra suggested as-grown graphene of high quality [22]. The transmission electron microscopy (TEM) image in Figure 1d clearly indicates edges of monolayer graphene, consistent with the Raman spectrum.

The flexibility and transparency of a monolayer graphene heater sample with an active heating area of ~$1.5 \times 1.5$ cm$^2$ is shown in Figure 2a. For comparison, a three-layer graphene heater is shown on the same white background, where it can be seen that the transparency changed in the center area. A transmittance of ~97.7% was reported for monolayer graphene [26]. The corresponding transmittances were ~95.4% for bilayer graphene, ~92.7% for three-layer graphene, ~90% for four-layer graphene, and ~87.3% for five-layer graphene. As shown in Figure 2b, the uniform surface temperature distribution, as obtained by infrared imaging, indicated a uniform graphene film well in line with the SEM image in Figure 1b. The resistance of a set of few-layer graphene-based heaters was evaluated using four-point probing. As shown in Figure 3, the monolayer resistance was close to 5 kΩ, while the resistance of the two- to five-layer graphene heaters was in the 1–1.5-kΩ range. The resistances of the four and five-layer graphene-based heaters appeared to be larger than that of the three-layer graphene heater, which could possibly be explained by the uncertainty of the transfer process causing wrinkles or cracks in the graphene film, or from PMMA residues left from the wet transfer process despite a careful rinse process being used.

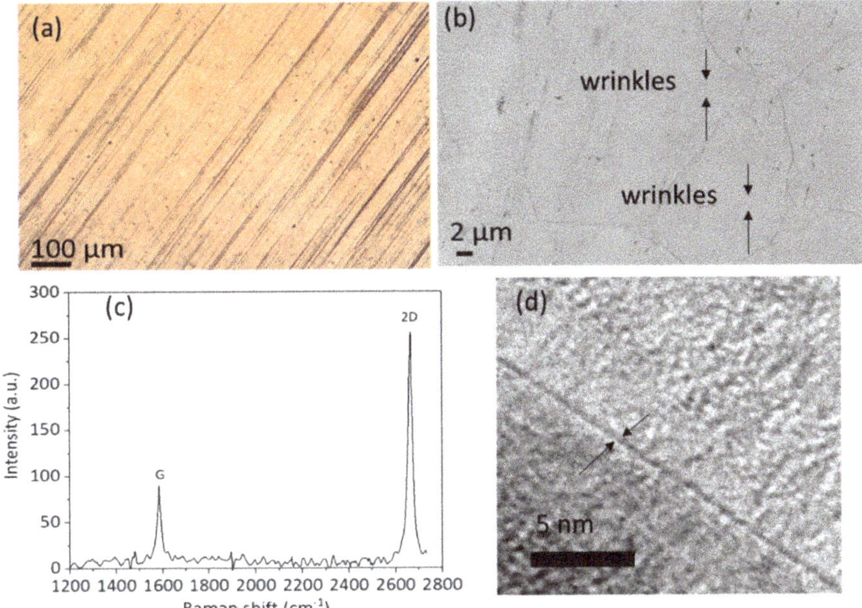


**Figure 1.** (**a**) Optical image of the morphology of Cu foil after graphene growth. (**b**) SEM image of the morphology of Cu foil after graphene growth. (**c**) Representative Raman spectrum (632 nm) of the graphene grown on Cu foil. (**d**) TEM images of synthesized graphene on Cu foil.

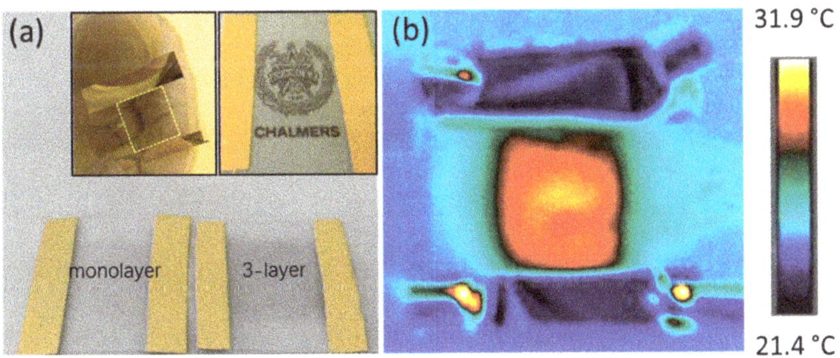


**Figure 2.** (**a**) Optical images of monolayer and three-layer graphene-based heater (top left inset: a monolayer graphene-based heater for illustrating flexibility; top right inset: a monolayer graphene-based heater placed on a Chalmers logo to illustrate transparency). (**b**) Typical infrared image showing the temperature distribution across the surface of a monolayer graphene heater. The infrared camera used was a high-resolution FLIR A655sc featuring a 640 × 480 pixel microbolometer that can detect temperature differences down to less than 30 mK.

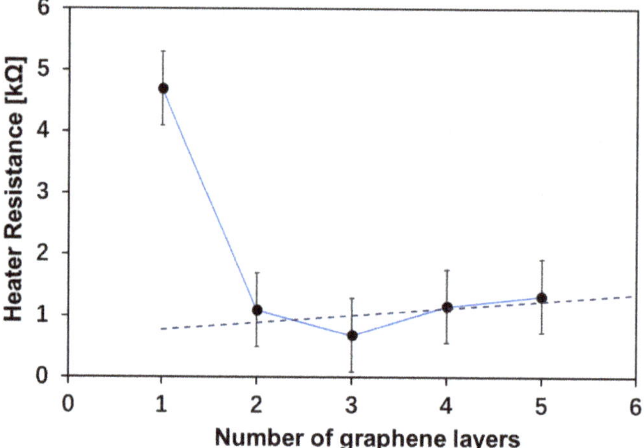


**Figure 3.** The resistance of the graphene-based heaters versus the number of graphene layers measured using the four-point probe method.

## 3. Electrothermal Performance of Graphene-Based Heaters

The heating mechanism of the graphene heaters was Joule heating. The electrothermal performance of monolayer graphene-based heaters determined using T-type thermocouples is shown in Figure 4, showing the heating and cooling behavior for six different values of applied input power. As shown in the figure, the heating and cooling behavior of the graphene-based heaters showed an exponential time dependence with a thermal time constant of 7 s. This time constant indicated the elapsed time required for the temperature difference of the graphene heater to rise to 63% of its final value during heating or, correspondingly to decay to 37% during cooling. It can also be interpreted as the time it would have taken to reach the final value if the heating/cooling continued at its initial rate. The heating and cooling behavior of the graphene heaters with other numbers of graphene layers showed a similar exponential time dependence, but with different time constants. As an example, the time constant for five-layer graphene heaters was 14 s during cooling and 10 s during heating. This difference between the heating and cooling rates may be attributed to the temperature-dependent electrical conductivity of graphene [27]. The model equation during cooling can be written as follows:

$$T = RT + \Delta T e^{-(t-t_0)/\tau}, \tag{1}$$

where RT is the room temperature, ΔT the temperature difference between room temperature and steady-state temperature, $\tau$ is the time constant, and $t_0$ is the time when the cooling starts.

From these experimental plots of the heating/cooling behavior of the graphene heaters, we could also extract the steady-state temperatures versus the input heating power. As shown in Figure 4, a steady-state temperature of 38 °C was obtained for an input power of 170 mW, while a steady-state temperature of 80 °C was obtained for an input power of 780 mW. The results are shown in Figure 5, where the steady-state temperatures were plotted as a function of the power density obtained by dividing the electrical input power by the 2.25-$cm^2$ area of the graphene heater. The choice of plotting versus the power density was done to enable a comparison between the performance of our graphene heater and other heaters previously reported in the literature. The plot in Figure 5 shows that graphene-based heaters, for the same input power density, reach much higher temperatures than, for instance, metal-based heaters [8,23]. For low temperatures and limited input power, heaters based on single-walled carbon nanotubes (CNTs) appear comparable, but there are no data available for higher temperatures [28]. Although details of the experimental set-ups may be different, it is obvious that the electrothermal performance of the graphene heaters in this study is much better than the performance of metal-based heaters and somewhat better than that of CNT-based heaters. The electrothermal performance of a laser-reduced graphene oxide heater [17] was even better than this work. It also seems to be a general trend that nanoscale carbon-based heaters electrothermally outperform heaters based on metallic films. However, the most important feature of the graphene-based heaters, making it worthwhile to investigate their performance for potential use in future applications, is their transparency.

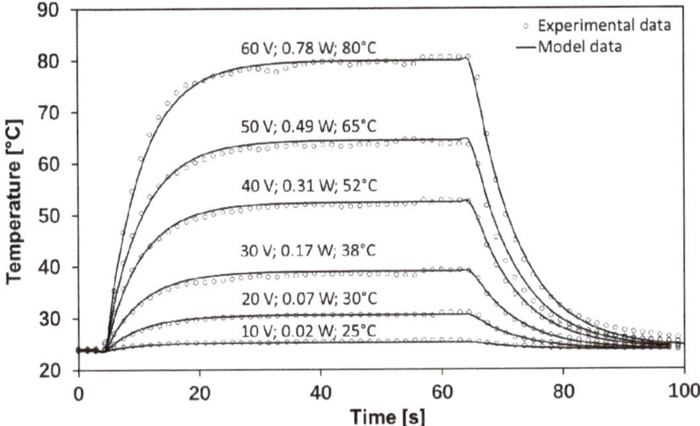


**Figure 4.** Time dependency of the electrothermal performance of a monolayer graphene-based heater for six different applied input voltages. The numbers occurring next to curves denote applied voltage, input power, and steady-state temperature, respectively. The temperature was logged using T-type thermocouples attached to the back side of the polyethylene terephthalate (PET) substrate.

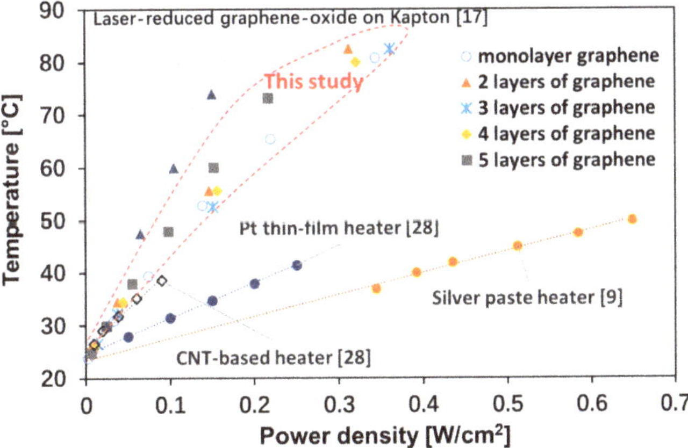


**Figure 5.** Steady-state temperatures vs. dissipated power density for graphene-based heaters with one, two, three, four, and five layers of graphene. Also shown for comparison are the same temperature versus power density relationships for two metallic heaters and one heater based on carbon nanotube films [9,17,28].

From the data in Figure 5, the convective heat-transfer coefficient, $h$, of the graphene heaters could be determined from the trendline slopes of the data for each type of graphene heater, with a calculation method similar to a previous report [4]. Theoretically, the steady-state temperature of a heater is determined by a balance between the electrical input power and the heat loss (mainly due to radiation and convection). For graphene, with its low emissivity, we can neglect the radiation loss and express the convection heat loss by the following equation:

$$Q/A = h\,\Delta T, \tag{2}$$

where Q/A is the input power density, and $\Delta T = T - T_0$ is the temperature difference between the steady-state temperature, $T$, and the ambient temperature, $T_0$. The resulting heat transfer coefficients are shown in Figure 6. There is no obvious trend for the dependence of the heat-transfer coefficient on the number of graphene layers, except possibly for the five-layer graphene heater that showed a somewhat lower value. For this reason, in Figure 5, only an average trendline is shown for the one- to four-layer graphene-based heaters ($h \approx 60$ W·m$^{-2}$·°C$^{-1}$). Similarly, the Pt thin-film heater had a heat transfer coefficient of ~150 W·m$^{-2}$·°C$^{-1}$, while the corresponding value for the silver paste heater was ~240 W·m$^{-2}$·°C$^{-1}$.

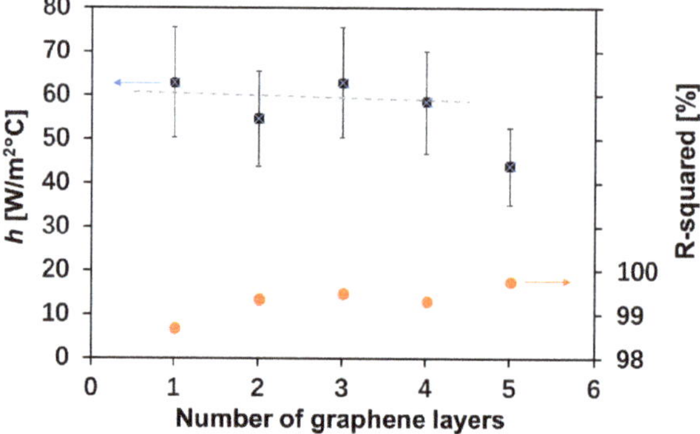


**Figure 6.** Convective heat-transfer coefficients for graphene-based heaters with from one to five layers of graphene. Also shown are the $R^2$ regression numbers of the trendline approximations.

The heat-transfer coefficients obtained for the graphene-based heaters were higher than the coefficient for natural convection of air (~5–25 W·m$^{-2}$·°C$^{-1}$), but in the range for that of forced convection of air (~20–200 W·m$^{-2}$·°C$^{-1}$). The higher value obtained here may be due to our experiments being conducted in a fume cupboard. Deviations between graphene samples were not appreciable, thereby validating the reliability of our measurements. The differences in heat-transfer coefficients between carbon-based heaters and metal thin-film heaters may be attributed to differences in the thermal interface conductance between the solid-gas adsorbates.

Finite element models were developed in COMSOL for studying the steady-state properties of graphene-based heaters and their surface temperature distributions. The electrically generated heat was modeled by using the electric currents and layered shell interface aimed at computing currents and potential distributions in thin conducting layers. Simulations using COMSOL for modeling properties of graphene were reported in previous work [29]. In this work, great care was taken to design the geometry of the model so that it would match the experimental behavior. For different electrical input power and the corresponding potential distributions across the heater surface, simulations resulted in an elevated temperature distribution across the surface of the PET substrate. For comparison between simulations and experiments, where the substrate temperatures were measured by a thermocouple, the average temperature of the backside of PET substrate at different input power densities was also calculated. A comparison between simulations and experiments is shown in Figure 7, where the simulation results show good agreement with the experimental results. What is not visible from the graph in Figure 7 is that, for the same power density, only half the applied voltage was needed for the five-layer graphene-based heater compared to the monolayer heater for the same input power due to the factor of four in their resistance difference. Moreover, the simulated temperature distribution is shown in Figure 8, where it can be seen that the steady-state surface temperature increased with the input voltage. Other graphene heaters had the same trend, whereby the steady-state surface temperature increased with the input voltage, but the values were different.

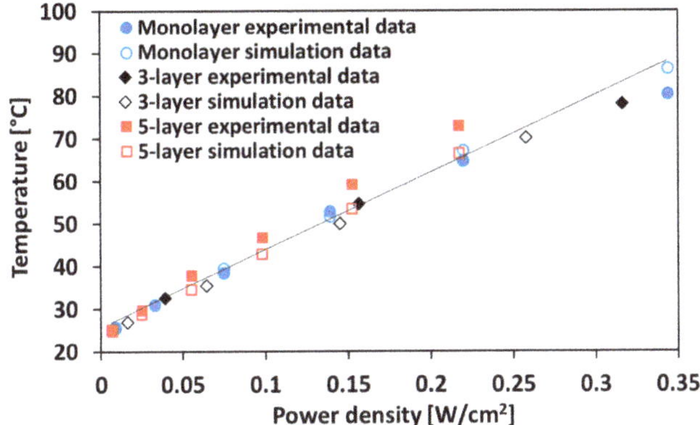


**Figure 7.** Simulated (open symbols) and experimental (solid symbols) steady-state surface temperatures for monolayer, three-layer, and five-layer graphene-based heaters.

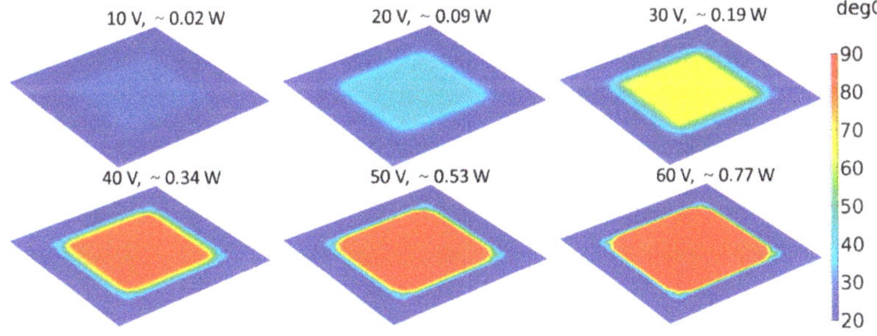


**Figure 8.** Steady-state surface temperature distributions of monolayer graphene-based heaters for six different applied voltages (the applied powers were also calculated).

Finally, an interesting observation made during our experiments was that the resistance of the graphene-based heaters increased after some time of exposure to air. We found that the resistance of the transferred graphene films was significantly increased after a one-year exposure to air, with increases for some samples as much as three to five times, which is much larger than reported in a previous study [30]. Other studies also reported that the resistance of transferred graphene films increased after storage in a humid environment [31,32]. As previously discussed, cracks and residues cannot be completely avoided during the transfer process from the copper foil on which the graphene is grown to the transparent PET substrate. Therefore, there is a risk that water molecules permeate the cracks and traverse the residues, which may weaken the graphene adhesion to the PET substrate and cause an increase in the resistances. It was also found that, when the PET substrates were placed on ice, the resistances of the graphene film increased by 10–20%. However, the resistance returned to the initial values after drying. The increase in resistance indicates that the adhesion between the graphene and the substrate plays a key role in the electrothermal performance of graphene-based heaters in real-world applications.

## 4. Conclusions

Based on the assumption that graphene, with its excellent electrothermal properties, could be a perfect material for transparent heaters in many applications, we designed and fabricated a set of graphene-based heaters with from one to five layers of CVD graphene grown on copper and transferred to a PET substrate. The properties of these graphene-based heaters were evaluated both experimentally and theoretically by simulations in terms of steady-state temperatures and in terms of heating/cooling rates versus the applied power density. In conclusion, we find our results promising in terms of quantifiable parameters such as thermal time constants, maximum heating/cooling rates, and convective heat-transfer coefficients when compared to Ag and Pt metal-based thin-film heaters. However, much work remains to refine the fabrication process and to improve the quality of the graphene films. As an example, we found that the quality of the graphene film and its adhesion to the PET substrate play a key role in determining the performance and reliability of graphene-based heaters when it comes to their electrothermal properties. The results presented here in this study strengthen our belief in graphene-based heaters as being promising candidates for the next generation of transparent heaters for various applications, such as anti-fog windows, mirror defoggers, and outdoor displays.

**Author Contributions:** Y.Z. conceptualized and designed the experiments, conducted the experiments together with H.L., and wrote the original draft manuscript. L.T. and Y.Z. conducted the simulations together with Y.Z., who also performed the data analysis together with K.J. The final manuscript was written by K.J. and Y.Z. All authors contributed to the discussion and analysis of the results regarding the final manuscript. All authors have read and agreed to the published version of the manuscript.

**Funding:** The authors acknowledge the financial support by the Key R & D Development Program from the Ministry of Science and Technology of China with the contract No. 2017YFB0406000, the Science and Technology Commission of Shanghai Municipality Program (19DZ2281000), and the National Natural Science Foundation of China (No. 51872182, No. 11672171, No. 11974236, and No. 61775130). J.L. also acknowledges the financial support from the Swedish Board for Innovation (Vinnova) under the Sioagrafen program, from the Swedish Board for Strategic Research (SSF) with the contract No. SE13-0061 and GMT14-0045, from Formas with the contract No. FR-2017/0009, and from the Swedish National Science Foundation with the contract No. 621-2007-4660, as well as from the Production Area of Advance at Chalmers University of Technology, Sweden.

**Conflicts of Interest:** The authors declare no conflicts of interest.

## References

1. Mo, Y.W.; Okawa, Y.; Tajima, M.; Nakai, T.; Yoshiike, N.; Natukawa, K. Micro-machined gas sensor array based on metal film micro-heater. *Sens. Actuat. B Chem.* **2001**, *79*, 175–181. [CrossRef]
2. Liu, P.; Liu, L.A.; Jiang, K.L.; Fan, S.S. Carbon-Nanotube-film microheater on a polyethylene terephthalate substrate and its application in thermochromic displays. *Small* **2011**, *7*, 732–736. [CrossRef]
3. Bae, J.J.; Lim, S.C.; Han, G.H.; Jo, Y.W.; Doung, D.L.; Kim, E.S.; Chae, S.J.; Huy, T.Q.; Luan, N.V.; Lee, Y.H. Heat dissipation of transparent graphene defoggers. *Adv. Funct. Mater.* **2012**, *22*, 4819–4826. [CrossRef]
4. Kiruthika, S.; Gupta, R.; Kulkarni, G.U. Large area defrosting windows based on electrothermal heating of highly conducting and transmitting Ag wire mesh. *RSC Adv.* **2014**, *4*, 49745–49751. [CrossRef]
5. Chuang, M.J. ITO films prepared by long-throw magnetron sputtering without oxygen partial pressure. *J. Mater. Sci. Technol.* **2010**, *26*, 577–583. [CrossRef]
6. De, S.; Higgins, T.M.; Lyons, P.E.; Doherty, E.M.; Nirmalraj, P.N.; Blau, W.J.; Boland, J.J.; Coleman, J.N. Silver nanowire networks as flexible, transparent, conducting films: Extremely high DC to optical conductivity ratios. *ACS Nano* **2009**, *3*, 1767–1774. [CrossRef]
7. Celle, C.; Mayousse, C.; Moreau, E.; Basti, H.; Carella, A.; Simonato, J.P. Highly flexible transparent film heaters based on random networks of silver nanowires. *Nano Res.* **2012**, *5*, 427–433. [CrossRef]
8. Ji, S.L.; He, W.W.; Wang, K.; Ran, Y.X.; Ye, C.H. Thermal response of transparent silver nanowire/PEDOT:PSS film heaters. *Small* **2014**, *10*, 4951–4960. [CrossRef] [PubMed]
9. Yoon, Y.H.; Song, J.W.; Kim, D.; Kim, J.; Park, J.K.; Oh, S.K.; Han, C.S. Transparent film heater using single-walled carbon nanotubes. *Adv. Mater.* **2007**, *19*, 4284–4287. [CrossRef]
10. Jang, H.S.; Jeon, S.K.; Nahm, S.H. The manufacture of a transparent film heater by spinning multi-walled carbon nanotubes. *Carbon* **2011**, *49*, 111–116. [CrossRef]

11. Kim, D.; Zhu, L.J.; Jeong, D.J.; Chun, K.; Bang, Y.Y.; Kim, S.R.; Kim, J.H.; Oh, S.K. Transparent flexible heater based on hybrid of carbon nanotubes and silver nanowires. *Carbon* **2013**, *63*, 530–536. [CrossRef]
12. Bolotin, K.I.; Sikes, K.J.; Jiang, Z.; Klima, M.; Fudenberg, G.; Hone, J.; Kim, P.; Stormer, H.L. Ultrahigh electron mobility in suspended graphene. *Solid State Commun.* **2008**, *146*, 351–355. [CrossRef]
13. Prasher, R. Graphene spreads the heat. *Science* **2010**, *328*, 185–186. [CrossRef] [PubMed]
14. Pang, S.P.; Hernandez, Y.; Feng, X.L.; Mullen, K. Graphene as transparent electrode material for organic electronics. *Adv. Mater.* **2011**, *23*, 2779–2795. [CrossRef] [PubMed]
15. Cao, Y.; Fatemi, V.; Fang, S.; Watanabe, K.; Taniguchi, T.; Kaxiras, E.; Jarillo-Herrero, P. Unconventional superconductivity in magic-angle graphene superlattices. *Nature* **2018**, *556*, 43–50. [CrossRef] [PubMed]
16. Sui, D.; Huang, Y.; Huang, L.; Liang, J.J.; Ma, Y.F.; Chen, Y.S. Flexible and transparent electrothermal film heaters based on graphene materials. *Small* **2011**, *7*, 3186–3192. [CrossRef] [PubMed]
17. Romero, F.J.; Rivadeneyra, A.; Ortiz-Gomez, I.; Salinas, A.; Godoy, A.; Morales, D.P.; Rodriguez, N. Inexpensive graphene oxide heaters lithographed by laser. *Nanomaterials* **2019**, *9*, 1184. [CrossRef]
18. Bobinger, M.R.; Romero, F.J.; Salinas-Castillo, A.; Becherer, M.; Lugli, P.; Morales, D.P.; Rodriguez, N.; Rivadeneyra, A. Flexible and robust laser-induced graphene heaters photothermally scribed on bare polyimide substrates. *Carbon* **2019**, *144*, 116–126. [CrossRef]
19. Menzel, R.; Barg, S.; Miranda, M.; Anthony, D.B.; Bawaked, S.M.; Mokhtar, M.; Al-Thabaiti, S.A.; Basahel, S.N.; Saiz, E.; Shaffer, M.S.P. Joule heating characteristics of emulsion-templated graphene aerogels. *Adv. Funct. Mater.* **2015**, *25*, 28–35. [CrossRef]
20. Bae, S.; Kim, H.; Lee, Y.; Xu, X.; Park, J.S.; Zheng, Y.; Balakrishnan, J.; Lei, T.; Kim, H.R.; Song, Y.I.; et al. Roll-to-roll production of 30-inch graphene films for transparent electrodes. *Nat. Nanotechnol.* **2010**, *5*, 574–578. [CrossRef]
21. Kang, J.; Kim, H.; Kim, K.S.; Lee, S.K.; Bae, S.; Ahn, J.H.; Kim, Y.J.; Choi, J.B.; Hong, B.H. High-performance graphene-based transparent flexible heaters. *Nano Lett.* **2011**, *11*, 5154–5158. [CrossRef] [PubMed]
22. Gao, Z.; Zhang, Y.; Fu, Y.; Yuen, M.M.F.; Liu, J. Thermal chemical vapor deposition grown graphene heat spreader for thermal management of hot spots. *Carbon* **2013**, *61*, 342–348. [CrossRef]
23. Zhang, Y.; Fu, Y.; Edwards, M.; Jeppson, K.; Ye, L.; Liu, J. Chemical vapor deposition grown graphene on Cu-Pt alloys. *Mater. Lett.* **2017**, *193*, 255–258. [CrossRef]
24. Li, X.; Cai, W.; An, J.; Kim, S.; Nah, J.; Yang, D.; Piner, R.; Velamakanni, A.; Jung, I.; Tutuc, E.; et al. Large-area synthesis of high-quality and uniform graphene films on copper foils. *Science* **2009**, *324*, 1312–1314. [CrossRef] [PubMed]
25. Ferrari, A.C.; Meyer, J.C.; Scardaci, V.; Casiraghi, C.; Lazzeri, M.; Mauri, F.; Piscanec, S.; Jiang, D.; Novoselov, K.S.; Roth, S.; et al. Raman spectrum of graphene and graphene layers. *Phys. Rev. Lett.* **2006**, *97*, 187401. [CrossRef] [PubMed]
26. Nair, R.R.; Blake, P.; Grigorenko, A.N.; Novoselov, K.S.; Booth, T.J.; Stauber, T.; Peres, N.M.R.; Geim, A.K. Fine structure constant defines visual transparency of graphene. *Science* **2008**, *320*, 1308. [CrossRef]
27. Fang, X.Y.; Yu, X.X.; Zheng, H.M.; Jin, H.B.; Wang, L.; Cao, M.S. Temperature- and thickness-dependent electrical conductivity of few-layer graphene and graphene nanosheets. *Phys. Lett. A* **2015**, *379*, 2245–2251. [CrossRef]
28. Kang, T.J.; Kim, T.; Seo, S.M.; Park, Y.J.; Kim, Y.H. Thickness-dependent thermal resistance of a transparent glass heater with a single-walled carbon nanotube coating. *Carbon* **2011**, *49*, 1087–1093. [CrossRef]
29. Subrina, S.; Kotchetkov, D.; Balandin, A.A. Heat removal in silicon-on-insulator integrated circuits with graphene lateral heat spreaders. *IEEE Electr. Device Lett.* **2009**, *30*, 1281–1283. [CrossRef]
30. Tan, L.F.; Zeng, M.Q.; Wu, Q.; Chen, L.F.; Wang, J.; Zhang, T.; Eckert, J.; Rummeli, M.H.; Fu, L. Direct growth of ultrafast transparent single-layer graphene defoggers. *Small* **2015**, *11*, 1840–1846. [CrossRef]
31. Hwang, J.; Kim, M.; Campbell, D.; Alsalman, H.A.; Kwak, J.Y.; Shivaraman, S.; Woll, A.R.; Singh, A.K.; Hennig, R.G.; Gorantla, S.; et al. van der Waals epitaxial growth of graphene on sapphire by chemical vapor deposition without a metal catalyst. *ACS Nano* **2013**, *7*, 385–395. [CrossRef] [PubMed]
32. Jung, W.; Park, J.; Yoon, T.; Kim, T.S.; Kim, S.; Han, C.S. Prevention of water permeation by strong adhesion between graphene and $SiO_2$ substrate. *Small* **2014**, *10*, 1704–1711. [CrossRef] [PubMed]

*Review*

# Thermal Characterization of Low-Dimensional Materials by Resistance Thermometers

**Yifeng Fu [1,*], Guofeng Cui [2] and Kjell Jeppson [1]**

1 Electronics Materials and Systems Laboratory, Department of Microtechnology and Nanoscience, Chalmers University of Technology, SE-41296 Gothenburg, Sweden; kjell.jeppson@chalmers.se
2 Key Laboratory for Polymeric Composite & Functional Materials of Ministry of Education, The Key Lab of Low-Carbon Chemistry and Energy Conservation of Guangdong Province, Materials Science Institute, School of Chemistry, Sun Yat-sen University, Guangzhou 510275, China; cuigf@mail.sysu.edu.cn
* Correspondence: yifeng.fu@chalmers.se; Tel.: +46-704-438-784

Received: 30 April 2019; Accepted: 24 May 2019; Published: 29 May 2019

**Abstract:** The design, fabrication, and use of a hotspot-producing and temperature-sensing resistance thermometer for evaluating the thermal properties of low-dimensional materials are described in this paper. The materials that are characterized include one-dimensional (1D) carbon nanotubes, and two-dimensional (2D) graphene and boron nitride films. The excellent thermal performance of these materials shows great potential for cooling electronic devices and systems such as in three-dimensional (3D) integrated chip-stacks, power amplifiers, and light-emitting diodes. The thermometers are designed to be serpentine-shaped platinum resistors serving both as hotspots and temperature sensors. By using these thermometers, the thermal performance of the abovementioned emerging low-dimensional materials was evaluated with high accuracy.

**Keywords:** thermal characterization; resistance temperature detector; heat spreader; carbon nanotube; graphene; boron nitride

---

## 1. Introduction

The semiconductor industry is pursuing electronic systems with higher integration density, more functions, higher power and frequency, and smaller footprint and volume, with lower cost. When the performance increases, the power density in electronics systems becomes higher and higher; thus, heat dissipation becomes a critical issue. In addition, the increase of hotspots and packaging complexity, such as in three dimensional (3D) stacking of processor and memory chips, makes thermal management an even more difficult task in microsystems. Various advanced materials and technologies were proposed and demonstrated to improve thermal management in electronics, for instance, nanoparticles and graphene-enhanced thermal interface materials (TIMs) [1], carbon nanotube (CNT)-based TIMs [2], cooling fins [3], etc. Therefore, thermal characterization of such nanomaterials and nanostructures becomes more important than ever to evaluate their performance.

Various methods were developed to characterize the thermal performance of nanomaterials. For instance, the thermal bridge method can be used to measure the in-plane thermal conductivity of extremely small structures down to a single atom layer [4]; the e-beam self-heating method can be used to measure the contact thermal resistance at material interfaces [5]; scanning thermal microscopy is able to map local temperature with nanoscale resolution and thermal conduction in materials [6]; the optothermal Raman spectroscopy technique allows high accuracy measurement of the thermal conductivity of atomic thick nanomaterials [7]; the pulsed photothermal reflectance method can be used to measure both thermal conductivity of materials and contact thermal resistance at interfaces [8]; the 3$\omega$ method allows high accuracy measurement of the thermal conductivity of materials [9]; the transient plane source method allows fast measurement of thermal conductivity, thermal diffusivity, and specific heat capacity of materials [10]; the laser flash method is also an easy-to-implement method for thermal conductivity measurement of materials [11]. It should be noted that the method should be selected depending on the size, geometry, composition, and performance of the materials in order to perform a proper characterization. Among all the thermal characterization methods, the on-chip resistance thermometer is a component allowing high accuracy, high speed, and real-time characterization of nanomaterials and nanostructures. This paper is expanded from a conference paper [12] but elaborates upon and includes the most recent published results to review the previous work on thermal characterization of various one- and two-dimensional (1D and 2D) nanomaterial-based cooling structures using resistance thermometers. First of all, the design, fabrication, and calibration of the resistance thermometer is presented. Secondly, we summarize the thermal characterizations of different low-dimensional materials using the resistance thermometer. This includes CNT-based cooling fins, graphene-based lateral heat spreaders, and boron nitride (BN)-based heat spreaders.

## 2. Resistance Thermometers

The principle of a resistance thermometer is to use temperature-sensitive materials to detect temperature by monitoring the change in electrical resistance of the material. Among all the materials, platinum (Pt) is one of the most used due to its highly linear temperature–resistance relationship. Fu et al. fabricated a resistance thermometer using e-beam evaporated Pt thin films on silicon chips [13], as shown in Figure 1. In order to realize the temperature monitoring in an embedded interface, they used through-silicon via (TSV) technology to read out the temperature. The serpentine Pt temperature sensors can also simultaneously act as heating elements to simulate hotspots in chips for the thermal characterization of heat dissipation materials and structures. After the fabrication, the resistance thermometers were calibrated by a standard resistance temperature detector (RTD). After calibration, the resistance thermometers can be used to monitor the temperature distribution on the thermal test chip; therefore, the cooling performance can be easily evaluated by simply measuring the resistance.

**Figure 1.** The thermal test chip with resistance thermometers and heating elements.

The thermal test chip fabricated by Fu et al. [13] shown in Figure 1 consists of a 3 × 3 array of thermometers with a size of 390 × 400 $\mu m^2$. The thickness of the platinum thermometers is 40 nm. Prior to the deposition of the platinum resistors, a 20-nm-thick titanium layer was deposited as an adhesion layer. The thickness of the insulating silicon dioxide ($SiO_2$) layer on the silicon substrate was 300 nm. Balandin et al. used a similar structure to model the heat spreading from metal–oxide–semiconductor (MOS) field-effect transistors on silicon-on-insulator (SOI) substrates with and without graphene heat spreaders [14].

In this paper, the thermal test chip shown in Figure 1 and its slightly modified version (to provide even higher power density) were used to evaluate the cooling performance of various nanomaterials and nanostructures, and the results are presented in Sections 3–5.

## 3. CNT-Based Micro Heat Sinks

Owing to the very strong $sp^2$ hybridized C–C bonding, CNTs exhibit excellent thermal properties. Therefore, they were proposed as a candidate for thermal interface material development and many results were reported [15–19]. On the other hand, since CNTs are mechanically strong [20,21] and can be vertically aligned, they can also be applied as heat sinks. CNT-based micro heat sinks were demonstrated to cool down power transistors by Mo et al. [3]. They grew CNTs on a silicon chip (as shown in Figure 2) and fabricated the cooler separately before attaching it onto the power transistor. It was found that the CNT-based cooler was able to cool down the power transistor to a much lower temperature (108 °C vs. 119 °C) even at much higher power input (25.7 W vs. 19.6 W). Fu et al. modified the design and fabricated the CNT cooling fins directly on top of the hotspots on silicon chips in order to further decrease the thermal resistance on the heat dissipation path [22], as shown in Figure 3. They firstly grew the CNT structures on a silicon substrate using Fe as a catalyst, and then transferred the CNT cooling fins onto a thermal test chip with high-power-density hotspots. Low-melting-point metal indium was used as the transfer media so that the transfer process would be compatible with complementary metal–oxide–semiconductor (CMOS) processes. The CNT fin structure was electrically insulated from the hotspot resistor by a 300-nm $SiO_2$ insulating layer on the hotspot circuit. More details about the transfer process can be found elsewhere [23,24]. Prior to the fabrication of the CNT cooling fins, multi-scale modeling was performed to optimize the dimension of the CNT structures (i.e., height, width, and pitch of the CNT fins); therefore, optimal pressure decrease (between coolant inlet and outlet) and maximal cooling effect were obtained.

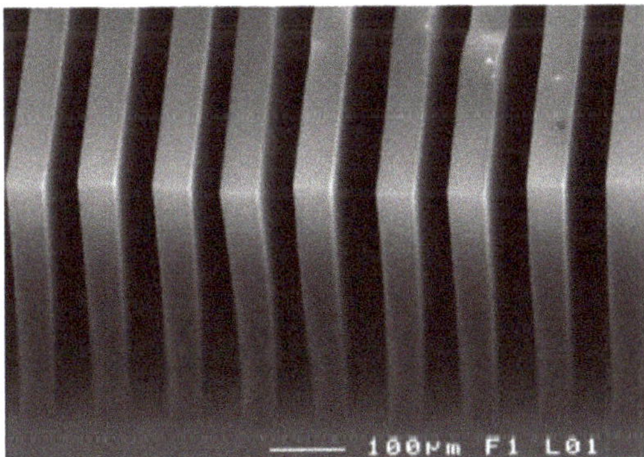


**Figure 2.** As-grown CNT cooling fins used to cool down the power transistor. Reprinted with permission from [3].

**Figure 3.** Transferred CNT cooling fins directly fabricated on top of the hotspot test structure. Reprinted with permission from Reference [22].

After transfer, the on-chip CNT-based micro heat sink was mounted onto a supporting circuit board as shown in Figure 4. To complete the cooling system, inlet and outlet nozzles were fabricated and connected to the CNT cooling fin structures through aluminum chambers at two ends of the test chip. Finally, polydimethylsiloxane (PDMS) was used to encapsulate the whole system to prevent coolant leakage. As a reference for studying the cooling performance of the CNT-based micro heat sink, identical cooling systems without CNT cooling fins were also fabricated and characterized.

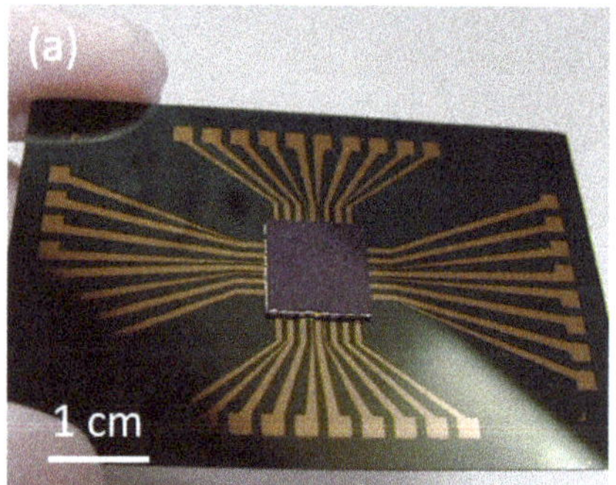

**Figure 4.** (**a**) On-chip CNT cooling fin test structure mounted on a supporting circuit board. (**b**) Complete cooling system embedded in polydimethylsiloxane (PDMS) with inlet and outlet nozzles for the coolant.

In order to examine the cooling performance of the CNT-based micro heat sink, air and water were used as coolant, and they were pumped to flow through the micro channels between the CNT fins. Some results of the experiments are shown in Figure 5 where the temperature at the hotspot is plotted vs. heat flux through the resistive hotspot.

As expected, water is a much more effective coolant than air. For a heat flux of 3000 W/cm$^2$, the hotspot temperature decreased by almost 50 °C (from 116 to 68 °C) upon using water at a flow rate of 0.32 m/s, compared to when air was used as the coolant, even though the air flow rate was ten times larger (3.2 m/s). However, more interesting is the unfortunate fact that the CNT cooling fins seemed to have a minimal influence when air was used as coolant. This is believed to be a combination of the thermal contact resistance to the hotspot being too high due to the interface layers, and that macro-scale cooling may not be directly scalable to a micro-scale environment.

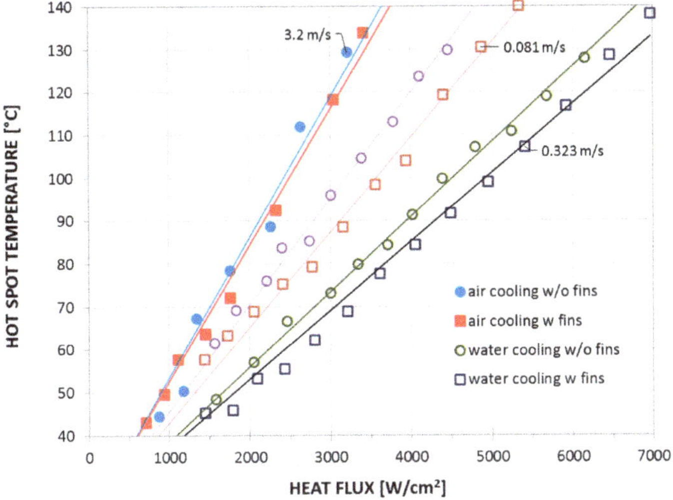


**Figure 5.** Cooling performance of the CNT-based micro heat sink using water as coolant (with air cooling as a reference) plotted as the hotspot temperature vs. heat flux. Experimental data sourced from Reference [22].

The experiments showed that, when the chip was cooled by water at a flow rate of 0.32 m/s, the hotspot temperature on the chip with the CNT cooling fin structure was about 8–10 °C lower than on the test chip without the CNT fins. Interestingly enough, beyond a certain flow rate of the water coolant, the cooling effect seems to be more or less independent of the flow rate, as shown in Figure 6. Since the water cooling of the indium adhesive seems to be so effective, the influence of the CNT cooling fins even appears to decrease as the flow rate of the water coolant increases beyond 0.08 m/s.

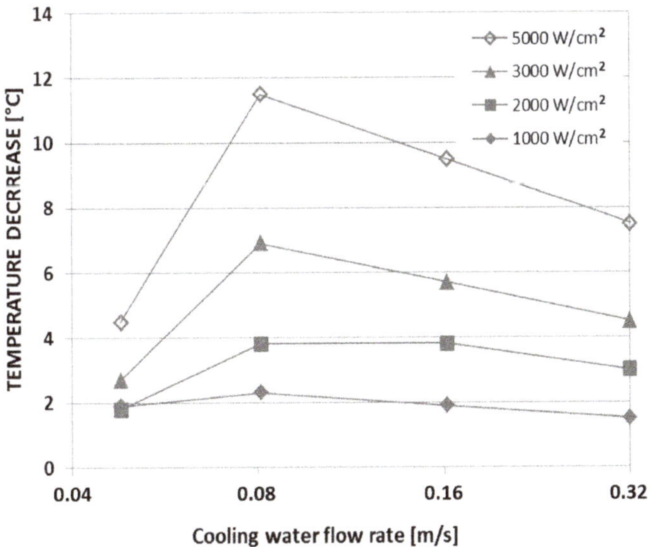


**Figure 6.** Hotspot temperature decrease vs. flow rate of water coolant for four different heat fluxes through the hotspot resistor. Experimental data sourced from Reference [22].

Finally, Figure 7 shows that the decrease of the hotspot temperature due to water cooling of the CNT fins seems to increase linearly with the flow rate of the water coolant.

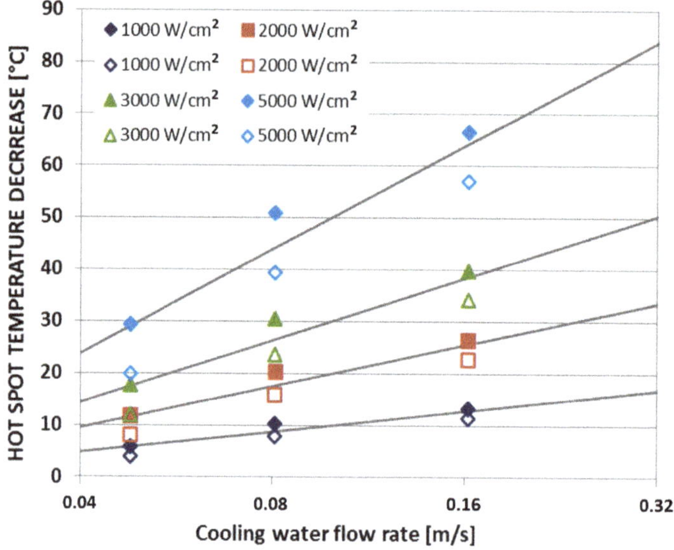


**Figure 7.** Hotspot temperature decrease vs. water coolant flow rate (test structure with CNT cooling fins = filled markers; test structure without CNT cooling fins = open markers). Experimental data sourced from Reference [22].

## 4. Graphene-Based Heat Spreaders

Similar to CNTs, graphene also possesses excellent thermal and mechanical properties due to its special crystalline structure [25]. In electronic systems, non-uniform distribution of thermal energy dissipates from high-power components, such as high-power transistors and light-emitting diodes (LEDs), leading to the formation of hotspots, together with high average device temperatures, resulting in the degradation of device performance and poor reliability. Therefore, various thermal composites [26–31] and heat spreaders [8,32–38] were developed and demonstrated using liquid-phase exfoliated (LPE) graphene and chemical vapor deposition (CVD)-grown graphene.

Balandin et al. showed that a few-layer graphene-based heat spreader connected to the drain of gallium nitride (GaN) high-power field-effect transistors considerably reduced the device temperature [32]. Using micro-Raman spectroscopy for in situ monitoring, they demonstrated that hotspot temperatures could be lowered by ~20 °C in transistors operating at a power density of ~13 W per mm of channel width, which they claim corresponds to an order-of-magnitude increase in device lifetime. Similarly, Hong et al. showed improved heat dissipation in gallium nitride LEDs by embedding reduced graphene oxide (rGO) patterns into the devices [33]. The infrared images of the LED chip surfaces from their paper shown in Figure 8 indicate a decrease in peak temperature on the chip surface from 58 °C for a conventional LED to 53 °C for the rGO-embedded LEDs. In addition, the average temperature on the chip surface decreased from 51 to 47 °C.

To evaluate the graphene-based heat spreaders, a new version of the resistance thermometer was designed and fabricated. Based on the lessons learnt from the CNT-based micro heat sink, the wires connecting the hotspot resistor and the I/O pads were redesigned to minimize the power dissipation via interconnect circuit. Two examples of such redesigned resistance thermometers are shown in Figure 9.

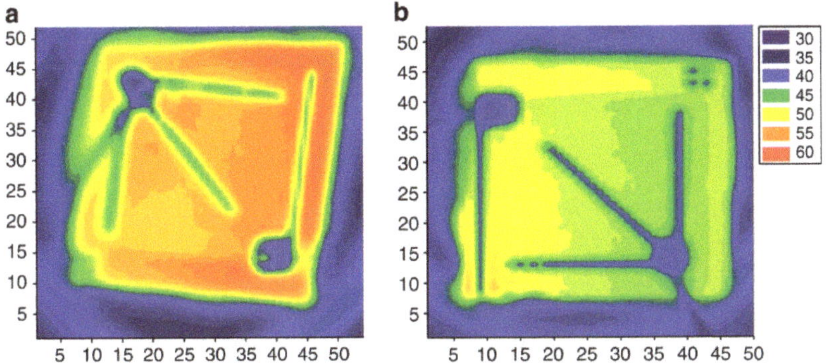


**Figure 8.** Infrared thermal imaging camera photographs of the chip surfaces showing the temperature distribution on the surface of (**a**) a conventional light-emitting diode (LED) chip, and (**b**) a reduced graphene oxide (rGO)-embedded chip under 100 mA current injections. Sourced from Reference [33]. Reprinted with permission from Nature Communications.

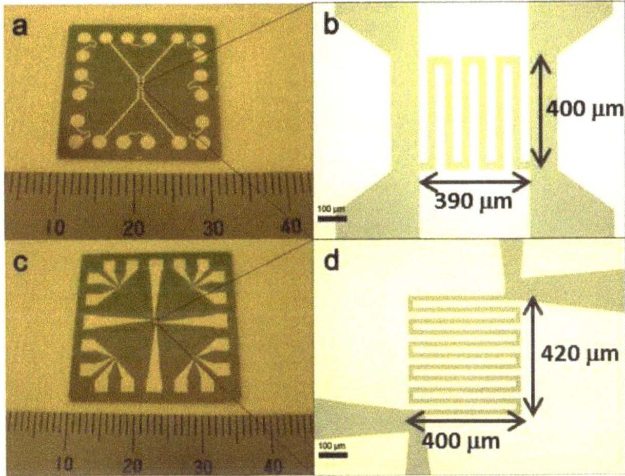


**Figure 9.** Redesigned resistance thermometers with larger area available for the heat spreader (**a**,**c**), and wider terminal wires (**b**,**d**).

These hotspot test structures were used in a series of experiments to investigate the thermal performance of 2D materials with high thermal conductivity, such as monolayer and multilayer graphene, and BN-based heat spreaders. By using such 2D materials as heat spreaders to dissipate the Joule heat generated from the hotspot laterally across the chip surface, both the hotspot temperature and the average temperature across the chip can be lowered. The area of the hotspot resistor used in these experiments was 390 × 400 µm$^2$, and its resistance was about 80 Ω at room temperature. monolayer graphene grown by chemical vapor deposition (CVD) was placed on the thermal test chip as heat spreader via the transfer method [34,35]. The graphene was isolated from the resistor by a 30 nm $SiO_2$ protective layer. Figure 10 shows the temperature vs. heat flux at the hotspot. It can be seen that the hotspot temperature can be decreased by about 10 °C by the graphene-based heat spreader (from 133 °C to 123 °C) at a heat flux of 460 W/cm$^2$. Thick graphene-based films fabricated from the liquid-phase exfoliation method [8,36,37] were also applied as heat spreaders in the same way as the monolayer graphene as shown in Figure 11. To decrease the thermal contact resistance, the thickness of the $SiO_2$ layer was reduced to one-tenth of the thickness that was used in the CNT cooling fin experiments. The detailed process of transferring and placing the monolayer and multilayer graphene heat spreader onto the hotspot structure is described elsewhere [34].

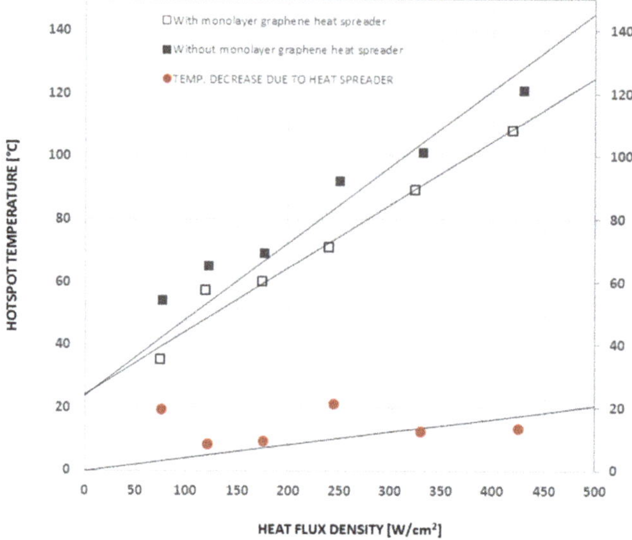


**Figure 10.** Temperature monitoring on thermal test chip with and without a monolayer graphene heat spreader. Replotted data sourced from Reference [34].

**Figure 11.** Multilayer transferred graphene film placed on hotspot test structure as a heat spreader across the chip surface.

In another investigation, an infrared camera was used to monitor the temperature on the thermal test chip to evaluate the cooling performance of a graphene-based heat spreader [38]. The thermal images in Figure 12 show the temperature distributions across the surface of the thermal test chip, which indicate that the temperature decreased by 5 °C when monolayer graphene was used as lateral heat spreader.

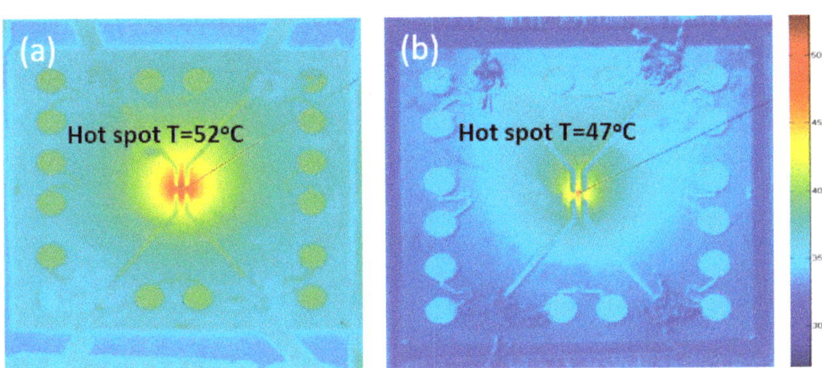


**Figure 12.** Temperature distributions on thermal test chips at a heat flux of 1280 W/cm$^2$ without (**a**) and with (**b**) a graphene heat spreader. Sourced from Reference [38].

A recent study showed that the cooling performance of a graphene-based heat spreader (fabricated via the vacuum filtration method) can be further improved by interfacial functionalization [36]. In a series of experiments, the graphene films were functionalized by (3-amino-propyl)-triethoxysilane (APTES) molecules to decrease the thermal contact resistance between the graphene-based heat spreader and the hotspot test structure. In this series of experiments, the redesigned resistance thermometer from Figure 9c was used.

The resulting thermal performance of the graphene-based heat spreader before and after functionalization is shown in Figure 13. It can be seen that the hotspot temperature on a bare chip without graphene heat spreader was 146 °C under a heat flux of 1500 W/cm$^2$. By placing a graphene film without functionalization on the surface of the test structure and repeating the measurements, the hotspot temperature was found to decrease to 140 °C ($\Delta T = 6$ °C). The estimated accuracy was ±0.5 °C. If, instead, the functionalized graphene-based heat spreader was used, where the thermal contact resistance between the graphene-based film and the test structure was reduced by the addition of a functionalized graphene oxide (FGO) interfacial layer, the hotspot temperature was found to decrease to 134 °C ($\Delta T = 12$ °C).

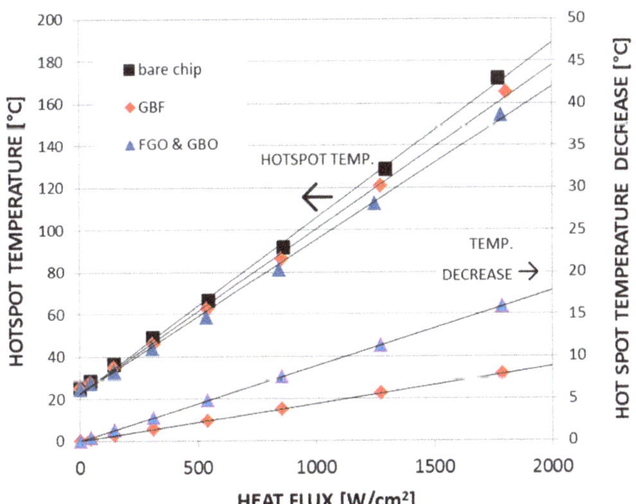


**Figure 13.** Cooling performance of functionalized graphene-based heat spreaders. Replotted data sourced from Reference [36].

## 5. Hexagonal Boron Nitride Heat Spreaders

In this paper, we also summarize the use of hotspot test structures for the evaluation of the performance of 2D hexagonal boron nitride (hBN) films as heat spreaders. The advantage of BN films over graphene is that they are electrically insulating and yet good thermal conductors [39]. In scenarios where electrical conduction is not allowed, hBN will be a very good complementary material to graphene for heat spreaders.

Bulk hBN has a typical thermal conductivity of 390 W/mK, which is 280 times higher than the thermal conductivity of silicon dioxide ($SiO_2$) insulators. For hBN monolayers, the thermal conductivity value can be even higher [40–42]. Thus, the advantage of hBN films is that they might be integrated to the semiconductor circuitry and be placed directly on top or below the hotspot without any insulating $SiO_2$ layers, which will significantly decrease the total thermal resistance along the heat conduction path and, therefore, greatly improve the cooling performance. For thermal management applications, 2D hBN was used to develop both thermal composites [43–45] and heat spreaders [46–48].

In the experiments to be summarized here, hBN films were transferred from the original growth substrate to the hotspot test structure via a similar method as the graphene films [32]. This transfer process includes spin-coating the hBN film with a supporting layer of polymethyl methacrylate (PMMA). The original growth substrate (Cu) was then etched away in a 30% $FeCl_3$ solution, leaving the PMMA-supported hBN film floating in $FeCl_3$ solution. The monolayer hBN film could then be transferred onto the calibrated hotspot test structure, before the PMMA was dissolved in hot acetone.

It should be noted that it is very challenging to fabricate freestanding pure hBN films since they are too brittle. Recently, Sun et al. successfully developed a process to fabricate flexible and uniform hBN films by adding acetate cellulose to the hBN dispersion [46]. Before thermal characterization on the hotspot test structure, the quality of the hBN material was examined by TEM. Results showed that few-layer hBN flakes were dominant in the film. The hotspot structure with an hBN heat spreader is shown in Figure 14. Thermal characterization was performed to evaluate the cooling performance of the hBN heat spreader using an infrared camera. Results showed that the hBN heat spreader can lower the hotspot temperature by almost 20 °C under a power density of 625 W/cm$^2$.

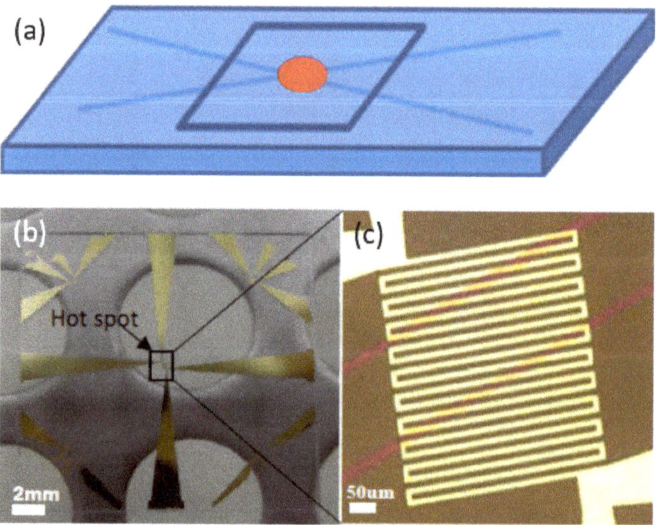


**Figure 14.** The hotspot test structure with the hexagonal boron nitride (hBN) heat spreader film. Reprinted with permission from [46].

In parallel to this study, Bao et al. applied monolayer hBN films as a lateral heat spreader to cool down the hotspot structure, as shown in Figure 9a [47]. Results showed that the performance of the monolayer hBN heat spreader on the hotspot fabricated on silicon substrates was not as good as in the case of the hotspot fabricated on quartz substrates. This is because a big portion of the heat was conducted through the Si substrate due to its higher thermal conductivity than quartz. Figure 15 shows the hotspot temperature under different power densities. It can be seen that, at a heat flux of 625 W/cm$^2$, the hotspot temperature can be reduced by 5 °C. When the heat flux was 1000 W/cm$^2$, the hotspot temperature could be reduced by 8 °C using the hBN heat spreader.

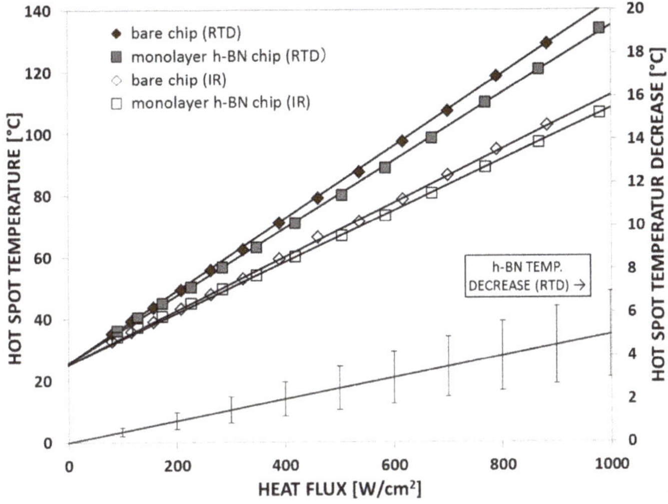


**Figure 15.** Cooling efficiency of the monolayer hBN heat spreader. Replotted data sourced from Reference [47].

Figure 15 also shows the temperature right below the hotspot (backside of the chip) measured by infrared (IR) camera. An example of such an IR image showing the temperature distribution on the backside of the chip is shown in Figure 16b. This photo again highlights the importance of a proper design of the test structure. The non-negligible resistance of the wires connecting the hotspot resistor with the output pads results in an asymmetrical temperature distribution due to the non-negligible power dissipated in the wires. The temperature distribution can be compared to the one obtained from the improved test structure design used in the previously described experiments. For the IR image captured from the front side of the test chip shown in Figure 16a, which was redesigned with appropriate wire widths and very low power dissipation through the connecting wires, the temperature distribution on the test chip was circular symmetric, which makes it easier to compare with a symmetrical simulation model.

For comparison, a similar study was performed using few-layer hBN films obtained from liquid-phase exfoliation (LPE) [48]. In this study, suspension of 2D hBN flakes was prepared with the assistance of sonication in an aqueous surfactant solution containing ethanol. The LPE process lasted 4 h and was followed by 20 min of centrifugation to get rid of the large BN particles. Afterward, the hBN suspension was drop-coated onto the hotspot test structure and then placed on a 60 °C hot plate to evaporate the solvent and obtain the multilayer hBN film as a lateral heat spreader. Details of the fabrication steps can be found in Reference [48]. The results of this study are shown in Figure 17. This graph shows the hotspot temperature vs. power density for three different samples. It can be seen that the temperature decrease at the hotspot was about 3–4 °C at a heat flux of 1000 W/cm$^2$—a result somewhat lower than that obtained for monolayer hBN films.

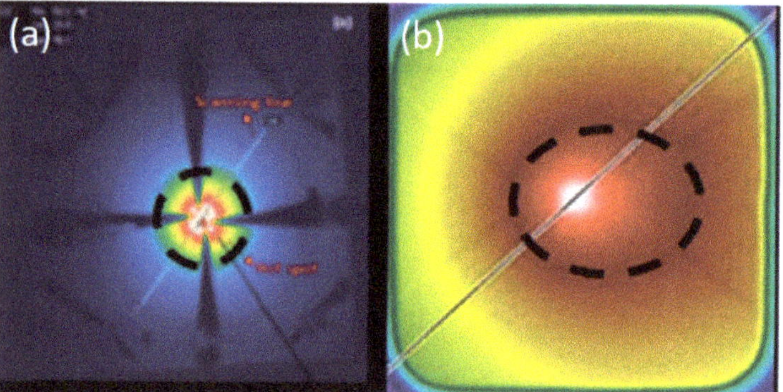


**Figure 16.** Temperature distribution across the hotspot test chip as captured by infrared camera for two different test structure designs: (**a**) new design; (**b**) old design. Reprinted with permission from Reference [46].

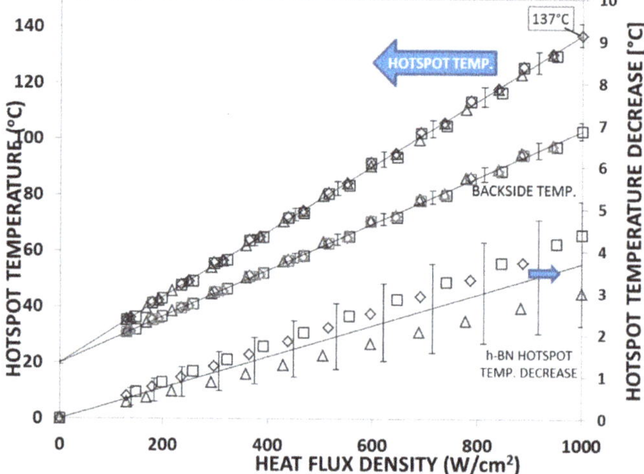


**Figure 17.** Bare chip hotspot temperature vs. power density by electrical and infrared measurements, as well as the hotspot temperature decrease due to the hBN heat spreader. Replotted data sourced from Reference [48].

It should be noted that there is larger variation in thermal performance between different hBN films (multilayer hBN films) as compared to variations between different monolayer hBN films. This is explained by the difficulties in maintaining the same properties between samples obtained by drop-coating of LPE hBN solutions. As shown in Reference [46], studies were also performed where the LPE hBN solution was enhanced by the addition of graphene.

## 6. Summary and Conclusion

A few emerging low-dimensional materials exhibit excellent thermal properties that could be used for thermal management of high-power electronics. In this paper, we reviewed a number of serpentine hotspot-producing and temperature-sensing test structures that can be used to evaluate the thermal performance of these 1D and 2D materials. The performances of both CNT-based micro heat sinks and two-dimensional films of graphene and hBN-based heat spreaders were summarized. For the CNT-based heat sink, air did not show much cooling effect, while water cooling could lower the hotspot temperature by 50 °C at high heat flux densities. Furthermore, several studies using monolayer graphene and hBN as a heat spreader were summarized. The monolayer graphene heat spreader was shown to be much more efficient in spreading the heat, thereby lowering the hotspot temperature, than the monolayer hBN heat spreader. Concerning few-layer graphene heat spreaders, it was shown that their performance could be improved considerably by functionalization using APTES, which can minimize the thermal contact resistance between the chip and the heat spreader. Few-layer hBN heat spreaders were shown to have similar heat spreading performance to few-layer graphene without functionalization (~5°C at 1000 W/cm$^2$).

These 1D and 2D materials show great potential as heat dissipation materials in electronics. However, challenges need to be addressed before the low-dimensional materials can be pushed onto the market. For the CNT-based micro heat sink, a CNT transfer process which can be upscaled to industry level and be compatible with the current semiconductor processes needs to be approved. For graphene-based heat spreaders, the thick graphene films are more favorable than the CVD-grown mono- to few-layer graphene films from the processability perspective. Lastly, hBN-based heat spreaders are easier to integrate into electronic systems than graphene-based heat spreaders because hBN films are electrically insulating; however, the mechanical strength of the hBN films needs to be improved.

**Author Contributions:** Writing—original draft preparation, Y.F. and K.J.; writing—review and editing, Y.F., G.C., and K.J.

**Funding:** This work was sponsored by the Swedish Foundation for Strategic Research (SSF) under contract Nos. SE13-0061 and GMT14-0045, and from the Nano Area of Advance (Dnr: C 2017-1256) and Production Area of Advance at Chalmers University of Technology, Sweden and EU Horizon 2020 projects Smartherm (690896) and Nanosmart (SEP-210506362). This work was also supported by the National Science Foundation of China under contract No. U1537104, the Shanghai Municipal Science and Technology Commission, and the Shanghai Municipal Education Commission (Shanghai University High Education Peak Discipline Program).

**Conflicts of Interest:** The authors declare no conflicts of interest.

## References

1. Shtein, M.; Nadiv, R.; Buzaglo, M.; Regev, O. Graphene-Based Hybrid Composites for Efficient Thermal Management of Electronic Devices. *ACS Appl. Mater. Interfaces.* **2015**, *7*, 23725–23730. [CrossRef] [PubMed]
2. Xu, J.; Fisher, T.S. Enhancement of thermal interface materials with carbon nanotube arrays. *Int. J. Heat Mass Transfer* **2006**, *49*, 1658–1666. [CrossRef]
3. Mo, Z.; Morjan, R.; Anderson, J.; Campbell, E.E.B.; Liu, J. Integrated nanotube microcooler for microelectronics applications. In Proceedings of the Electronic Components and Technology, Orlando, FL, UAS, 31 May–3 June 2005; pp. 51–54.
4. Kim, P.; Shi, L.; Majumdar, A.; McEuen, P.L. Thermal transport measurements of individual multiwalled nanotubes. *Phys. Rev. Lett.* **2001**, *87*, 215502-1. [CrossRef]
5. Wang, Z.; Xie, R.; Bui, C.T.; Liu, D.; Ni, X.; Li, B.; Thong, J.T.L. Thermal Transport in Suspended and Supported Few-Layer Graphene. *Nano Lett.* **2011**, *11*, 113–118. [CrossRef]
6. Choi, D.; Poudel, N.; Park, S.; Akinwande, D.; Cronin, S.B.; Watanabe, K.; Taniguchi, T.; Yao, Z.; Shi, L. Large Reduction of Hot Spot Temperature in Graphene Electronic Devices with Heat-Spreading Hexagonal Boron Nitride. *ACS Appl. Mater. Interfaces.* **2018**, *10*, 11101–11107. [CrossRef]
7. Balandin, A.A.; Ghosh, S.; Bao, W.; Calizo, I.; Teweldebrhan, D.; Miao, F.; Lau, C.N. Superior Thermal Conductivity of Single-Layer Graphene. *Nano Lett.* **2008**, *8*, 902–907. [CrossRef]
8. Han, H.; Zhang, Y.; Wang, N.; Samani, M.K.; Ni, Y.; Mijbil, Z.Y.; Edwards, M.; Xiong, S.; Sääskilahti, K.; Murugesan, M.; et al. Functionalization mediates heat transport in graphene nanoflakes. *Nat. Commun.* **2016**, *7*, 11281.
9. Chen, Z.; Jang, W.; Bao, W.; Lau, C.N.; Dames, C. Thermal contact resistance between graphene and silicon dioxide. *Appl. Phys. Lett.* **2009**, *95*, 161910. [CrossRef]

10. Ramirez, S.; Chan, K.; Hernandez, R.; Recinos, E.; Hernandez, E.; Salgado, R.; Khitun, A.G.; Garay, J.E.; Balandin, A.A. Thermal and magnetic properties of nanostructured densified ferrimagnetic composites with graphene - graphite fillers. *Mater. Des.* **2017**, *118*, 75–80. [CrossRef]
11. Jeon, D.; Kim, S.H.; Choi, W.; Byon, C. An experimental study on the thermal performance of cellulose-graphene-based thermal interface materials. *Int. J. Heat Mass Transf.* **2019**, *132*, 944–951. [CrossRef]
12. Jeppson, K.; Bao, J.; Huang, S.; Zhang, Y.; Sun, S.; Fu, Y.; Liu, J. Hotspot test structures for evaluating carbon nanotube microfin coolers and graphene-like heat spreaders. In Proceedings of the 2016 International Conference on Microelectronic Test Structures (ICMTS), Yokohama, Japan, 28–31 March 2016; pp. 32–36.
13. Fu, Y.; Wang, T.; Jonsson, O.; Liu, J. Application of through silicon via technology for in situ temperature monitoring on thermal interfaces. *J. Micromech. Microeng.* **2010**, *20*, 025027. [CrossRef]
14. Subrina, S.; Kotchetkov, D.; Balandin, A.A. Heat Removal in Silicon-on-Insulator Integrated Circuits With Graphene Lateral Heat Spreaders. *IEEE Electron Device Lett.* **2009**, *30*, 1281–1283. [CrossRef]
15. Nylander, A.N.; Fu, Y.; Huang, M.; Liu, J. Covalent Anchoring of Carbon Nanotube-Based Thermal Interface Materials Using Epoxy-Silane Monolayers. *IEEE Trans. Compon. Packag. Manuf. Technol.* **2019**, *9*, 427–432. [CrossRef]
16. Cross, R.; Cola, B.A.; Fisher, T.; Xu, X.; Gall, K.; Graham, S. A metallization and bonding approach for high performance carbon nanotube thermal interface materials. *Nanotechnology* **2010**, *21*, 445705. [CrossRef] [PubMed]
17. Huang, H.; Liu, C.H.; Wu, Y.; Fan, S. Aligned carbon nanotube composite films for thermal management. *Adv. Mater.* **2005**, *17*, 1652–1656. [CrossRef]
18. Biercuk, M.J.; Llaguno, M.C.; Radosavljevic, M.; Hyun, J.K.; Johnson, A.T.; Fischer, J.E. Carbon nanotube composites for thermal management. *Appl. Phys. Lett.* **2002**, *80*, 2767. [CrossRef]
19. Lin, W.; Moon, K.S.; Wong, C.P. A combined process of in situ functionalization and microwave treatment to achieve ultrasmall thermal expansion of aligned carbon Nanotube-Polymer nanocomposites: Toward applications as thermal interface materials. *Adv. Mater.* **2009**, *21*, 2421–2424. [CrossRef]
20. Lu, J.P. Elastic properties of carbon nanotubes and nanoropes. *Phys. Rev. Lett.* **1997**, *79*, 1297. [CrossRef]
21. Yu, M.F.; Lourie, O.; Dyer, M.J.; Moloni, K.; Kelly, T.F.; Ruoff, R.S. Strength and breaking mechanism of multiwalled carbon nanotubes under tensile load. *Science* **2000**, *287*, 637–640. [CrossRef]
22. Fu, Y.; Nabiollahi, N.; Wang, T.; Wang, S.; Hu, Z.; Carlberg, B.; Zhang, Y.; Wang, X.; Liu, J. A complete carbon-nanotube-based on-chip cooling solution with very high heat dissipation capacity. *Nanotechnology* **2012**, *23*, 045304. [CrossRef]
23. Fu, Y.; Qin, Y.; Wang, T.; Chen, S.; Liu, J. Ultrafast Transfer of Metal-Enhanced Carbon Nanotubes at Low Temperature for Large-Scale Electronics Assembly. *Adv. Mater.* **2010**, *22*, 5039–5042. [CrossRef]
24. Fu, Y.; Ye, L.L.; Liu, J. Thick film patterning by lift-off process using double-coated single photoresists. *Mater. Lett.* **2012**, *76*, 117–119. [CrossRef]
25. Balandin, A.A. Thermal properties of graphene and nanostructured carbon materials. *Nat. Mater.* **2011**, *10*, 569–581. [CrossRef]
26. Shahil, K.M.; Balandin, A.A. Graphene–Multilayer Graphene Nanocomposites as Highly Efficient Thermal Interface Materials. *Nano Lett.* **2012**, *12*, 861–867. [CrossRef]
27. Goli, P.; Legedza, S.; Dhar, A.; Salgado, R.; Renteria, J.; Balandin, A.A. Graphene-enhanced hybrid phase change materials for thermal management of Li-ion batteries. *J. Power Sources* **2014**, *248*, 37–43. [CrossRef]
28. Saadah, M.; Hernandez, E.; Balandin, A. Thermal Management of Concentrated Multi-Junction Solar Cells with Graphene-Enhanced Thermal Interface Materials. *Appl. Sci.* **2017**, *7*, 589. [CrossRef]
29. Shtein, M.; Nadiv, R.; Buzaglo, M.; Kahil, K.; Regev, O. Thermally Conductive Graphene-Polymer Composites: Size, Percolation, and Synergy Effects. *Chem. Mater.* **2015**, *27*, 2100–2106. [CrossRef]
30. Gu, J.; Xie, C.; Li, H.; Dang, J.; Geng, W.; Zhang, Q. Thermal percolation behavior of graphene nanoplatelets/polyphenylene sulfide thermal conductivity composites. *Polym. Compos.* **2014**, *35*, 1087–1092. [CrossRef]
31. Li, A.; Zhang, C.; Zhang, Y.F. Thermal Conductivity of Graphene-Polymer Composites: Mechanisms, Properties, and Applications. *Polymers* **2017**, *9*, 437.
32. Yan, Z.; Liu, G.; Khan, J.M.; Balandin, A.A. Graphene quilts for thermal management of high-power GaN transistors. *Nat. Commun.* **2012**, *3*, 827. [CrossRef] [PubMed]

33. Han, N.; Cuong, T.V.; Han, M.; Ryu, B.D.; Chandramohan, S.; Park, J.B.; Kang, J.H.; Park, Y.J.; Ko, K.B.; Kim, H.Y.; et al. Improved heat dissipation in gallium nitride light-emitting diodes with embedded graphene oxide pattern. *Nat. Commun.* **2013**, *4*, 1452. [CrossRef] [PubMed]
34. Gao, Z.; Zhang, Y.; Fu, Y.; Yuen, M.M.; Liu, J. Thermal chemical vapor deposition grown graphene heat spreader for thermal management of hot spots. *Carbon* **2013**, *61*, 342–348. [CrossRef]
35. Gao, Z.; Zhang, Y.; Fu, Y.; Yuen, M.; Liu, J. Graphene heat spreader for thermal management of hot spots in electronic packaging. In Proceedings of the 18th International Workshop on THERMal INvestigation of ICs and Systems, Budapest, Hungary, 25–27 September 2012; pp. 1–4.
36. Zhang, Y.; Han, H.; Wang, N.; Zhang, P.; Fu, Y.; Murugesan, M.; Edwards, M.; Jeppson, K.; Volz, S.; Liu, J. Improved Heat Spreading Performance of Functionalized Graphene in Microelectronic Device Application. *Adv. Funct. Mater.* **2015**, *25*, 4430–4435. [CrossRef]
37. Zhang, Y.; Edwards, M.; Samani, M.K.; Logothetis, N.; Ye, L.; Fu, Y.; Jeppson, K.; Liu, J. Characterization and simulation of liquid phase exfoliated graphene-based films for heat spreading applications. *Carbon* **2016**, *106*, 195–201. [CrossRef]
38. Huang, S.; Zhang, Y.; Sun, S.; Fan, X.; Wang, L.; Fu, Y.; Zhang, Y.; Liu, J. Characterization for graphene as heat spreader using thermal imaging method. In Proceedings of the 2013 14th International Conference on Electronic Packaging Technology, Dalian, China, 11–14 August 2013; pp. 403–408.
39. Zhou, H.; Zhu, J.; Liu, Z.; Yan, Z.; Fan, X.; Lin, J.; Wang, G.; Yan, Q.; Yu, T.; Ajayan, P.M.; et al. High thermal conductivity of suspended few-layer hexagonal boron nitride sheets. *Nano Res.* **2014**, *7*, 1232–1240. [CrossRef]
40. Lindsay, L.; Broido, D.A. Theory of thermal transport in multilayer hexagonal boron nitride and nanotubes. *Phys. Rev. B* **2012**, *85*, 035436. [CrossRef]
41. Ouyang, T.; Chen, Y.; Xie, Y.; Yang, K.; Bao, Z.; Zhong, J. Thermal transport in hexagonal boron nitride nanoribbons. *Nanotechnology* **2010**, *21*, 245701. [CrossRef]
42. Bao, J.; Jeppson, K.; Edwards, M.; Fu, Y.; Ye, L.; Lu, X.; Liu, J. Synthesis and applications of two-dimensional hexagonal boron nitride in electronics manufacturing. *Electron. Mater. Lett.* **2016**, *12*, 1–16. [CrossRef]
43. Wang, Z.; Iizuka, T.; Kozako, M.; Ohki, Y.; Tanaka, T. Development of epoxy/BN composites with high thermal conductivity and sufficient dielectric breakdown strength part I - sample preparations and thermal conductivity. *IEEE Trans. Dielectr. Electr. Insul.* **2011**, *18*, 1963–1972. [CrossRef]
44. Lei, Y.; Han, Z.; Ren, D.; Pan, H.; Xu, M.; Liu, X. Design of h-BN-Filled Cyanate/Epoxy Thermal Conductive Composite with Stable Dielectric Properties. *Macromol. Res.* **2018**, *26*, 602–608. [CrossRef]
45. Yang, N.; Zeng, X.; Lu, J.; Sun, R.; Wong, C.P. Effect of chemical functionalization on the thermal conductivity of 2D hexagonal boron nitride. *Appl. Phys. Lett.* **2018**, *113*, 171904. [CrossRef]
46. Sun, S.; Bao, J.; Mu, W.; Fu, Y.; Zhang, Y.; Ye, L.; Liu, J. Cooling hot spots by hexagonal boron nitride heat spreaders. In Proceedings of the 2015 IEEE 65th Electronic Components and Technology Conference (ECTC), San Diego, CA, USA, 26–29 May 2015; pp. 1658–1663.
47. Bao, J.; Zhang, Y.; Huang, S.; Sun, S.; Lu, X.; Fu, Y.; Liu, J. Application of two-dimensional layered hexagonal boron nitride in chip cooling. *J. Basic Sci. Eng.* **2016**, *24*, 210–217.
48. Bao, J.; Edwards, M.; Huang, S.; Zhang, Y.; Fu, Y.; Lu, X.; Yuan, Z.; Jeppson, K.; Liu, J. Two-dimensional hexagonal boron nitride as lateral heat spreader in electrically insulating packaging. *J. Phys. D: Appl. Phys.* **2016**, *49*, 265501. [CrossRef]

MDPI
St. Alban-Anlage 66
4052 Basel
Switzerland
Tel. +41 61 683 77 34
Fax +41 61 302 89 18
www.mdpi.com

*Materials* Editorial Office
E-mail: materials@mdpi.com
www.mdpi.com/journal/materials

www.ingramcontent.com/pod-product-compliance
Lightning Source LLC
LaVergne TN
LVHW070612170726
843515LV00004B/60

* 9 7 8 3 0 3 9 3 6 2 0 4 2 *